T0282702

The theme of this book is the investigation of globally asymptotic solutions of evolutionary equations. Locally asymptotic solutions of the Navier-Stokes equations and reaction-diffusion equations are the starting point, and by considering perturbed evolutionary equations, global approximations are constructed.

The lectures upon which this book is based were warmly received at the universities of Rome and Pavia, and at the Scuola Normale Superiore in Pisa. Here Professor Vishik has collated his lecture notes, and has added an appendix describing his recent work on attractors deriving from dynamical systems. This is unquestionably a fine addition to the Lezioni Lincee, and will be a necessary addition to the library of all who seek an insight into the solution of evolutionary equations.

Asymptotic behaviour of
solutions of evolutionary equations

Lezioni Lincee
Editor: Luigi A. Radicati di Brozolo, Scuola Normale Superiore, Pisa

This series of books arises from lectures given under the auspices of the Accademia Nazionale dei Lincei through a grant from IBM Italia.
The lectures, given by international authorities, will range on scientific topics from mathematics and physics through to biology and economics. The books are intended for a broad audience of graduate students and faculty members, and are meant to provide a *'mise au point'* for the subject they deal with.
The symbol of the Accademia, the Lynx, is noted for its sharp sightedness; the volumes in the series will be penetrating studies of scientific topics of contemporary interest.

Already published

Chaotic Evolution and Strange Attractors: D. Ruelle
Introduction to Polymer Dynamics: P. de Gennes
The Geometry and Physics of Knots: M. Atiyah
Attractors for Semigroups and Evolution Equations: O. Ladyzhenskaya

Asymptotic behaviour of solutions of evolutionary equations

M. I. VISHIK
Moscow State University

CAMBRIDGE
UNIVERSITY PRESS

CAMBRIDGE UNIVERSITY PRESS
Cambridge, New York, Melbourne, Madrid, Cape Town, Singapore,
São Paulo, Delhi, Dubai, Tokyo, Mexico City

Cambridge University Press
The Edinburgh Building, Cambridge CB2 8RU, UK

Published in the United States of America by Cambridge University Press, New York

www.cambridge.org
Information on this title: www.cambridge.org/9780521422376

First published 1992

A catalogue record for this publication is available from the British Library

ISBN 978-0-521-42023-5 Hardback
ISBN 978-0-521-42237-6 Paperback

To the memory of L. A. Lusternik

Contents

Introduction

This book is based on a series of lectures (the so called 'Lezioni Lincee') delivered by the author in the Accademia dei Lincei, Rome University, the Scuola Normale Superiore (Pisa) and Pavia University.

The main attention is given to the global asymptotics of solutions $u(t) = u(t, x)$ of evolutionary equations

$$\partial_t u = A(u), \quad u \mid_{t=0} = u_0. \tag{1}$$

The main results are illustrated by examples of the Navier-Stokes equations, the system of reaction-diffusion equations, the semilinear wave equation with dissipation, and other systems of parabolic and hyperbolic equations.

In the Appendix non-autonomous dynamical systems are studied. We assume that the dependence on time is quasi-periodic or almost periodic. We investigate problems related to the existence and dimension of attractors.

Let problem (1) in the autonomous case have a unique solution $u(t)$, $t \geq 0$, and let $u(t)$ belong to the same Banach space E for each t. In this case to system (1) there corresponds a semigroup of operators $\{S_t \mid t \geq 0\}$, $S_t : E \to E$, where $S_t u_0 = u(t)$ and $u(t)$ is the solution of (1). So we shall study the global behaviour of solutions mainly in terms of this semigroup.

The book includes the results obtained by the author in collaboration with A. V. Babin, M. Y. Skvortsov, V. Y. Skvortsov and V. V. Chepyzhov.

In the first chapter we remind the reader of the definitions relating to the maximal (global) attractor of a semigroup $\{S_t\}$ and formulate the existence theorem. We also give the definition of an invariant manifold containing an equilibrium point z of a semigroup $\{S_t\}$ and state its main properties.

In the second chapter the spectral asymptotics are described in the simplest case when the solution $u(t)$ of (1) tends to an equilibrium point z (not necessarily stable) as $t \to +\infty$.

We note that when equation (1) is linear and the solution can be found as a Fourier series, then the spectral asymptotics of $u(t)$ coincides with the partial sum $u_N(t)$ of its Fourier series. Let λ_k be the k-th eigenvalue of the operator A. Then the inequality

$$\|u(t) - u_N(t)\| \leq C e^{-\lambda_{N+1} t} \tag{2}$$

holds. For simplicity we assume that A is a self-adjoint operator semi-bounded from above and that it has a discrete spectrum $\{-\lambda_n\}$, where $\lambda_n \to +\infty$ for $n \to \infty$.

Asymptotics similar to (2) also exist when the operator $A(u)$ is nonlinear and $u(t) \to z$, $t \to +\infty$, where z is an equilibrium point of the semigroup $\{S_t\}$.

We assume that the operators S_t have the Fréchet derivative $S_t'(z)$ at the point z. Obviously, the set of the operators $S_t'(z)$ for $t \geq 0$ form a linear semigroup. If the circle $|\zeta| = \rho$ does not intersect the spectrum of the operator $S_1'(z)$, we call ρ a regular spectral radius. To the part of spectrum lying outside this circle there corresponds an invariant subspace of $S_1'(z)$ (it is also invariant with respect to $S_t'(z)$ for $t \geq 0$). We denote this space by $E_+(z, \rho)$ and assume that $E_+(z, \rho)$ is finite-dimensional. If some natural conditions are satisfied, then there exists a finite-dimensional $M_+(z, \rho)$, which is locally invariant with respect to $\{S_t\}$ and is tangent to $E_+(z, \rho)$ at the point z.

In §3 it is proved that if $u(t) \to z$ when $t \to +\infty$ and ρ is a regular spectral radius, then on the manifold $M_+(z, \rho)$ there exists a trajectory $\tilde{u}(t)$ (spectral asymptotic for $u(t)$) such that

$$\|u(t) - \tilde{u}(t)\| \leq C\rho^t \quad \forall t \geq T. \tag{3}$$

In the case of nonlinear parabolic equations or the Navier-Stokes system we can take a sequence of regular radii $\rho_j \to 0$, and then the corresponding sequence $\tilde{u}_j \in M_+(z, \rho_j)$ will approximate $u(t)$ with increasing accuracy (Babin & Vishik [1], [7]).

We note that the rate of approach of $u(t)$ to the equilibrium point z was studied by Foias & Saut [1], Foias & Guillopé [1], Haraux [1], [2], and other authors. The relationship between the trajectories belonging to $M_+(z, \rho)$ ($\rho = 1 - \varepsilon$) and arbitrary trajectories belonging to a neighbourhood of the point z was studied by Pliss [1]. These results were applied to the problem of stability or instability of the equilibrium point z of ordinary differential equations system.

In Chapter III uniform asymptotics (with respect to the initial data u_0, when u_0 belongs to a bounded set $B \subseteq E$) is constructed. It is assumed that the semigroup $\{S_t\}$ has a global Lyapunov function P, which decreases along all trajectories $S_t u_0$ except for the equilibrium points z. We also assume that the set $\mathcal{M}$ of equilibrium points z is finite, $\mathcal{M} = (z_1, ..., z_N)$. In this case uniform asymptotics of the trajectories $u(t)$ are described in terms of piecewise continuous trajectories lying on $\cup_j M^+(z_j, \rho_j)$, where $M^+(z_j, \rho_j) = \cup_{t \geq 0} S_t M_+(z_j, \rho_j)$. Each continuous part of the asymptotic $\tilde{u}(t)$ lies on one of the manifolds $M_+(z_j, \rho_j)$, and the values of $\tilde{u}(t)$ at its discontinunity points, lie in a small neighbourhood of an unstable equilibrium point $z_j \in \mathcal{M}$. The number of such points does not exceed N. Obviously, the trajectories $\tilde{u}(t)$ form a finitely-parametrized family of spectral asymptotics. In Chapter V for arbitrary trajectory $u(t) = S_t u_0$, $u_0 \in B$, a spectral asymptotic $\tilde{u}(t)$ is constructed such that

$$\|u(t) - \tilde{u}(t)\| \leq C e^{-\eta t} \quad \forall t \geq 0, \tag{4}$$

where the constant C depends only on B and $\eta > 0$ depends on spectral radii ρ_j ($j = 1, ..., N$). If all $\rho_j \to 0$, then $\eta \to +\infty$ (Babin & Vishik [1], [4], [7]).

The relationship between the above constructions and the ideas of Fourier, Lyapunov and Poincaré is quite clear.

We also note that although the dynamic system corresponding to the semigroup $\{S_t\}$ is considered in the infinite-dimensional space E, its asymptotics are described by means of a finite-dimensional dynamical system generated by the semigroup $\{S_t\}$ on the finite-dimensional manifolds $M^+(z_j, \rho_j)$.

In §§4 and 5 we give examples of equations and systems having local and uniform spectral asymptotics. Among these examples are various equations from mathematical physics such as the two- and three-dimensional Navier-Stokes systems, the reaction-diffusion system, quasilinear parabolic equations, and a hyperbolic equation with dissipation.

In §6 of Chapter IV we study perturbation theory for equations with a parameter,

$$\partial_t u = A(u, \lambda), \quad u_{t=0} = u_0 \quad (|\lambda| \leq \lambda_0). \tag{5}$$

Naturally, the corresponding semigroup also depends on λ, i.e. $S_t = S_t(\lambda)$.

A number of results on the behaviour of the solutions $u(t, \lambda)$ for large t were given by Hale [2], Mora & Sola-Morales [1] and other authors.

We state the following results.

Let $u = u(t, \lambda)$ be a solution to problem (5) and let $v(t) = S_t(0)u_0$ be the zeroth term of the λ-asymptotics, i.e., a solution of (5) for $\lambda = 0$. Then, under some natural conditions on $A(u, \lambda)$, the following estimate holds:

$$\|u(t, \lambda) - v(t)\| \leq C|\lambda|e^{\alpha t}, \quad \alpha > 0. \tag{6}$$

It is clear that with increasing t the estimate (6) worsens. Let us suppose that the operator $S_t(0)$ satisfies the conditions for the existence of uniform spectral asymptotics in the case when all $\rho_j = 1$ and the following inequality holds:

$$\|S_t(\lambda)u_0 - S_t(0)v_0\| \leq Ce^{\alpha t}(|\lambda|^\gamma + \|u_0 - v_0\|), \quad \gamma > 0. \tag{7}$$

In this case the above function $v(t) = S_t(0)u_0$ can be stabilised for $t \to +\infty$. More precisely, for each $u(t) = S_t(\lambda)u_0$, where $u_0 \in B$ (B is a bounded set in E), there exists a trajectory $\tilde{u}(t)$ such that

$$\tilde{u}(t) = v(t) = S_t(0)u_0$$

for $t \in [0, T(u)]$ and

$$\tilde{u}(t) \in \bigcup_j M^+(z_j, 1)$$

for $t > T$. As above, for $t > T$ the approximation $\tilde{u}(t)$ is piecewise continuous with respect to t, and $\tilde{u}(t)$ belongs to a finitely-parametrized family of trajectories $\{S_t(0)\}$. The continuous parts of $\tilde{u}(t)$ lie on the manifolds $M^+(z_j, 1)$. It will be shown (in §6) that $\tilde{u}(t)$ can be chosen so that the estimate

$$\sup_{t \geq 0} \|u(t, \lambda) - \tilde{u}(t)\| \leq C|\lambda|^q$$

holds uniformly with respect to u_0 when u_0 belongs to a bounded set B. Here $C = C(B)$ and $q > 0$ depends on the spectral characteristic of operators $S_1'(z_j)$ (Babin & Vishik [1], [4], [7]).

The stabilised asymptotics are illustrated in §7 by examples of the reaction-diffusion system, hyperbolic systems with dissipation, and parabolic systems of equations with a small parameter in spatial derivatives of the highest order.

In the second, third and fourth chapters we give a summary of some results obtained by Babin & Vishik [1], [4], [6], [7].

In Chapter V we describe stabilized asymptotics for the reaction-diffusion system with a small parameter in the time derivative:

$$\left.\begin{array}{l} \varepsilon\partial_t u_1 = \Delta u_1 - f_1(u_1, u_2) - g_1(x), \quad x \in \Omega \subseteq \mathbf{R}^n, \\[4pt] \partial_t u_2 = \Delta u_2 - f_2(u_1, u_2) - g_2(x), \\[4pt] u_1\mid_{\partial\Omega} = 0, \quad u_1\mid_{\partial\Omega} = 0 \text{ or } \dfrac{\partial u_1}{\partial\nu}\mid_{\partial\Omega} = 0, \quad \dfrac{\partial u_2}{\partial\nu}\mid_{\partial\Omega} = 0, \end{array}\right\} \quad (8)$$

and initial conditions

$$u_1\mid_{t=0} = u_1^0, \quad u_2\mid_{t=0} = u_2^0.$$

We note that when $\varepsilon = 0$, the first equation of this system becomes stationary. The semigroup $\{S_t(\varepsilon)\}$ that corresponds to system (8) acts in the phase space $E = H_1(\Omega) \times H_1(\Omega)$, and for $\varepsilon = 0$ it turns into the semigroup $\{S_t(0)\}$ which acts in another phase space $E_1 = H_1(\Omega)$. This action ($\varepsilon = 0$) can be determined, if we find $u_1 = u_1(u_2)$ from the first (stationary) equation (with $\varepsilon = 0$) in (8), substitute it into the second equation, and then find $S_t u_2(0) = u_2(t)$, where $u_2(t)$ is the solution of the second equation with the initial data $u_2(0)$.

When $f_1(u_1, u_2)$ and $f_2(u_1, u_2)$ satisfy some conditions and the initial functions are bounded, we can construct stabilised asymptotics

$$(v_1(t), v_2(t))$$

such that

$$\left.\begin{array}{l} \sup\limits_{t\geq 0}\|u_2(t,\varepsilon) - v_2(t)\| \leq C\varepsilon^q, \\[6pt] \sup\limits_{t\geq\tau}\|u_1(t,\varepsilon) - v_1(t)\| \leq C\varepsilon^q. \end{array}\right\} q > 0, \tau > 0. \quad (9)$$

Here the functions $v_1(t)$, $v_2(t)$ are piecewise continuous, each continuous part of $v_1(t)$, $v_2(t)$ is a solution of (8) for $\varepsilon = 0$, and each continuous part of $v_2(t)$ (except the first part) lies on a finite-dimensional unstable manifold of the semigroup $\{S_t(0)\}$ (Vishik & V. Skvortsov [1], [2]).

Problems with a small parameter in spatial derivatives of the highest order are studied in Chapters VI and VII.

A characteristic feature of these problems is the presence of a boundary layer term in the asymptotics. For simplicity, we consider a parabolic equation

$$\frac{\partial u}{\partial t} = -\varepsilon^2\Delta^2 u + \Delta u - f(u) - g(x), \quad (10)$$

$$u\mid_{\partial\Omega} = 0, \quad \frac{\partial u}{\partial\nu}\mid_{\partial\Omega} = 0, \quad x \in \Omega \subseteq \mathbf{R}^n. \quad (11)$$

For $\varepsilon = 0$ the equation (10) is a parabolic equation of the second order, and we can make it satisfy only the first boundary condition.

We study an asymptotic expansion of the elements $u(x)$ lying on the attractor $\mathcal{A}^\varepsilon$ for the problem (10), (11). Under some conditions on $f(u)$ the following expansion holds:

$$u(x) = (u_0 + v_0 - \varepsilon c_0) + \varepsilon(u_1 + v_1 - \varepsilon c_1) + r, \qquad (12)$$

where $u_0(x)$ and $u_1(x)$ have uniformly bounded norms with respect to $u(x) \in \mathcal{A}^\varepsilon$ in $H_3(\Omega)$. The functions v_0 and v_1 are of the form of a boundary layer, that is in the neighbourhood of $\partial\Omega$ we have $v_i(x) = \varepsilon c_i(x')e^{-\rho/\varepsilon}$, where $x' \in \partial\Omega$ and ρ is the distance along the normal at x' from this point to x and $c_0(x)$ and $c_1(x)$ are smooth functions.

The remainder r admits of the estimate

$$\|r\|_{H_3(\Omega)} + \varepsilon^{-1}\|r\|_{H_2(\Omega)} + \varepsilon^{-2}\|r\|_{H_1(\Omega)} \le M, \qquad (13)$$

where the constant M does not depend on ε and on $u \in \mathcal{A}^\varepsilon$ (M. Skvortsov & Vishik [1]). Let us note that the function $u(x)$ has a bounded norm only in $H_{3/2}(\Omega)$, and the index $3/2$ cannot be increased (see M. Skvortsov [1], [2], [3]). The 'boundary layer' functions v_0 and v_1 have bounded norms in $H_{3/2}(\Omega)$ uniformly with respect to ε for $0 < \varepsilon \le \varepsilon_0$. Obviously, these functions are quite essential in the expansion (12).

Formula (12) is similar to the asymptotic expansion of the solution of an elliptic equation with small parameter in the highest derivatives, which was studied by Vishik & Lusternik [1], [2].

Denote by $U(R)$ the set of trajectories $\{u(t)\}$ of equation (10) with bounded initial data, $\|u(t_0)\| \le R$, $t_0 \in \mathbf{R}$ fixed. For any element $u(T)$, for fixed $T > t_0$ and $u(t) \in U(R)$, the following asymptotic expansion is established:

$$u(T) = u_0 + v + r. \qquad (14)$$

Here the function u_0 is bounded in $H_3(\Omega)$ uniformly with respect to ε and for $u(t) \in U(R)$, v is the first-order boundary layer function (see Vishik & Lusternik [1], [2]), and r is the remainder.

From the expansion (14) we derive in §16 the asymptotics of the trajectory $u(t)$ on any finite time interval $t \in [t_1, t_1 + T_1]$, $t_1 \in \mathbf{R}$, $T_1 > 0$:

$$u(t) = w(t) + v(t) + r(t) \qquad (15)$$

(Vishik & M. Skvortsov [2]). Here $w(t)$ is the solution of the limit (for $\varepsilon = 0$) parabolic equation

$$\partial_t w = \Delta w - f(w) - g, \quad w\mid_{\partial\Omega} = 0 \qquad (16)$$

with initial data $w(t_1) = u_0$, where u_0 is defined in (14) for $T = t_1$, $v(t)$ is the boundary layer function of the first order, and $r(t)$ is the remainder.

Let the set $\mathcal{M} = (z_1, ..., z_N)$ of equilibrium points of (16) be finite. Let $\delta > 0$ and let the trajectory $u(t)$ belonging to $U(R)$ for $t \in (t_1, t_2)$ be outside the δ-neighbourhood $\mathcal{O}_\delta(\mathcal{M})$. Then under certain conditions on f and g, (i) the time $T = t_2 - t_1$, for which $u(t)$ is outside $O_\delta(\mathcal{M})$, is bounded uniformly for $\varepsilon \in (0, \varepsilon_0]$ and $u(t) \in U(R)$, (ii) the asymptotic expansion (15) of the trajectory $u(t)$ for $t \in [t_1, t_2]$ is valid and the remainder $r(t)$ satisfies the following inequality:

$$\varepsilon^2 ||r(t)||_2^2 + ||r(t)||_1^2 + \int_{t_1}^{t_2} ||\partial_t r||^2 \, d\tau \le C\varepsilon^2, \tag{17}$$

where $C = C(\delta, R)$ is independent of ε and of $u(t)$ (M. Vishik & M. Skvortsov [2]).

In §§15 and 17 similar results are established for the set of solutions of equation (10) with boundary conditions

$$u\,|_{\partial\Omega} = 0, \quad \Delta u\,|_{\partial\Omega} = 0. \tag{18}$$

The global asymptotics of the trajectories of problem (10), (18) for any $t \ge t_0$ is given. It is similar to (15) if $u(t) \notin \mathcal{O}_\delta(\mathcal{M})$, and estimate (17) holds. Inside the neighbourhood $\mathcal{O}_{\delta_0}^0(\mathcal{M})$ (in the metric of H) only the principal asymptotic term is constructed (the stabilized asymptotics as in Chapter IV). It consists of parts of trajectories $w(t)$ of the limit equation.

The results obtained in works by M. Skvortsov and M. Vishik are given in the sixth and seventh chapters.

In the Appendix we present the results of works of V. Chepyzhov and M. Vishik. It contains the study of non-autonomous infinite-dimensional dynamical systems generated by evolutionary equations:

$$\left.\begin{array}{l} \partial_t u = A(u, t), \quad t \ge \tau, \\ u\,|_{t=\tau} = u_\tau, \quad \tau \in \mathbf{R}, \quad u_\tau \in E. \end{array}\right\} \tag{19}$$

Here E is a Banach space. We assume that the nonlinear operator $A(u, t)$ depends on time, and that this dependence is quasi-periodic or almost periodic. We consider the following problems of mathematical physics as examples of (19).

1. The two-dimensional Navier-Stokes system:

$$\partial_t u = -vLu - B(u, u) + g(x, t), \tag{20}$$
$$u = (u^1, u^2), \quad u\,|_{\partial\Omega} = 0, \quad x \in \Omega \Subset \mathbf{R}^2,$$

where

$$L = \Pi\Delta, \quad B(u,u) = \Pi \sum_{i=1}^{2} \left(u^i \partial_{x_i} u^i \right)$$

where Π is the projection to the subspace of vector functions with a free divergence.

2. The reaction-diffusion system:

$$\partial_t u = a\Delta u - f(u,x,t) + g(x,t), \tag{21}$$

$$u \mid_{\partial\Omega} = 0 \text{ or } \frac{\partial u}{\partial v} \mid_{\partial\Omega} = 0,$$

where $u = (u^1, ..., u^n)$, $f = (f^1, ..., f^n)$, $g = (g^1, ..., g^n)$, $x \in \Omega \Subset \mathbf{R}^n$, and the matrix $a + a^* > 0$ and the function f satisfies some natural conditions.

3. The hyperbolic equation with dissipation:

$$\partial_t u^2 + \gamma \partial_t u = \Delta u - f(u,x,t) + g(x,t), \tag{22}$$

$$u \mid_{\partial\Omega} = 0, \quad \gamma > 0,$$

which occurs in quantum field theory.

We state some new results concerning the existence of attractors for the above problems and estimate the dimension of these attractors.

If problem (19) is well posed and $u(t) \in E$, $\forall t \geq \tau$, then the solution $u(t)$ can be represented in the form $u(t) = U(t,\tau)u_\tau$, where $U(t,\tau)$: $E \to E$, $t \geq \tau$, $\tau \in \mathbf{R}$. The family of maps $\{U(t,\tau) \mid E \to E, t \geq \tau, \tau \in \mathbf{R}\}$ is called the process on E generated by problem (19). By analogy with the autonomous case, we introduce the concepts of absorbing and attracting set for the process $\{U(t,\tau)\}$. The closed set $\mathcal{A}_1 \subseteq E$ is called the attractor of the process $\{U(t,\tau)\}$ if it is attracting (dist$_E(U(t,\tau)B, \mathcal{A}_1) \to 0$, $t - \tau \to +\infty$, where B is an arbitrary bounded set in E) and satisfies the minimal property. The minimal property is the generalization of the invariance property in the definition of an attractor of a semigroup. (See Sell [1], [2], Dafermos [2], [3], Haraux [3], where the main notions were introduced and some important results concerning attractors of processes were established).

In the Appendix we consider a family of processes $\{U_G(t,\tau)\}$ depending on some functional parameter G called the time symbol of a process. This parameter G is determined by the right-hand side g and, in examples (21), (22), by the nonlinear function f. We prove some general theorems on the existence of a uniform attractor of a family of processes with almost periodic time symbol. We describe a simple method for constructing such attractors. Let, for example, the operator $A(u,t)$ be quasi-periodic with respect to t, i.e.

$$A(u,t) = A_1(u,\omega_1,...,\omega_k) = A_1(x,\omega)$$

where A_1 is 2π-periodic with respect to each ω_i, $\omega = \Lambda t + \omega_0 \in T^k$ ($\omega_i = \lambda_i t + \omega_{i0}$, $i = 1,...,k$), $\Lambda = (\lambda_1,...,\lambda_k)$, $\lambda_1,...,\lambda_k$ are rationally independent real numbers, and T^k is a k-dimensional torus. Then equation (1) (or equations (20), (21), (22)) can be reduced to the autonomous system

$$\left.\begin{aligned}\partial_t u &= A_1(u,\omega), \quad u\,|_{t=0} = u_0, \\ \partial_t \omega &= \Lambda, \quad \omega\,|_{t=0} = \omega_0.\end{aligned}\right\} \tag{23}$$

System (23) generates the semigroup $\{S(t)\}$ acting on $E \times T^k$ according to the formula

$$S(t)(u_0,\omega_0) = (u(t), \Lambda t + \omega) \quad \forall t \geq 0,$$
$$S(t) : E \times T^k \quad \forall t \geq 0.$$

If the semigroup $\{S(t)\}$ has a compact absorbing or attracting set and is continuous, then the attractor $\mathcal{A} \in E \times T^k$ of the semigroup $\{S(t)\}$ exists. The projection $\Pi_1 \mathcal{A} = \mathcal{A}_1 \in E$, where Π_1 is the projection to the first component u of the pair (u,ω), is the uniform attractor with respect to $\omega_0 \in T^k$ of the process $\{U_{\omega_0}(t,\tau)\}$ generated by system (19) in the quasi-periodic case. It is shown in the Appendix that problems (20), (21), (22) possess uniform attractors $\mathcal{A}_1$ when the functions $g(x,t)$ and $f(u,x,t)$ are quasi-periodic or almost periodic.

Note that if $g(t) = g(x,t)$ is an almost periodic function, then to define the semigroup $\{S(t)\}$ one must use, instead of T^k, the hull $H(g)$ of the almost periodic function g. Problems (21) and (22) use analogous constructions for almost periodic functions $f(u,x,t)$.

Using this technique we obtain upper bounds for the Hausdorff dimension of attractors for problems (20) and (21) in the quasi-periodic case. For instance, we prove the following estimate for the Hausdorff dimension of the attractor $\mathcal{A}_1$ of the Navier-Stokes system (20):

$$\dim_H \mathcal{A}_1 \leq k + Ck^{1/3} + \mathcal{O}(1). \tag{24}$$

Here we assume, for simplicity, that the Reynolds numbers are bounded ($\nu = \mathcal{O}(1), \|g\| \leq C_1$). Thus, if the number k of rational independent frequencies is increasing, then the upper bound for the dimension of the attractor $\mathcal{A}_1$ also increases. Examples show that estimate (24) is exact.

In the last section of the Appendix we give upper bounds for the Hausdorff dimension of sections of attractors $\mathcal{A}_1$ when g and f are fixed and time t is also fixed.

Let $X = X_{g,f} = \{u(t)\}$ be a family of all bounded solutions for $t \in \mathbf{R}$ to the equation (20), (21) or (22) with given external force g and nonlinear function f (in the case of (20) or (21)). The section $X(t_0)$ at time t_0 is the union of all $u(t_0)$ for $u(t) \in X$. It is shown that the Hausdorff dimension of $X(t)$ is finite. For instance, for the Navier-Stokes system we have

$$\dim_H X(t_0) \leq \frac{C}{\nu^2} \quad \forall t_0 \in \mathbf{R}.$$

This estimate agrees with the corresponding result for the attractor dimension in the autonomous case.

I
Preliminaries

§1 Attractors of evolutionary equations

When we study a nonlinear evolutionary equation

$$\partial_t u = Au, \quad u\,|_{t=0} = u_0$$
$$(u = u(t,x), \quad u = (u_1, ..., u_N), \quad x = (x_1, ..., x_n)), \tag{1.1}$$

it is very important to understand the behaviour of its solutions when $t \to \infty$. If the problem (1.1) has a unique solution $u(t)$ for all $t \geq 0$ which belongs to some Banach space E, then to the problem corresponds a semigroup of operators $\{S_t, t \geq 0\}$ acting from E to E by the law

$$S_t u(t_1) = u(t + t_1), \quad t \geq 0, \quad t_1 \geq 0.$$

The action of the semigroup evidently satisfies the semigroup identity

$$S_t \cdot S_\tau = S_{t+\tau}. \tag{1.2}$$

The equation (1.1) is an infinite-dimensional dynamic system. Various examples of such systems (Navier-Stokes system, equation of reaction-diffusion, wave equation with dissipation and others) have been intensively studied in recent years. The main attention has been concentrated on the study of attractors of these equations.

We give necessary definitions. Let us denote by $\mathcal{B}(E)$ the set of all bounded sets belonging to E.

Definition 1.1.

A set $K \subseteq E$ is called an attracting set of the semigroup $\{S_t\}$ if for any $B \in \mathcal{B}(E)$

$$\mathrm{dist}_E(S_t B, K) \to 0 \quad \text{when} \quad t \to \infty.$$

Let us recall that for any $\mathcal{C} \subseteq E, \mathcal{D} \subseteq E$

$$\mathrm{dist}_E(\mathcal{C}, \mathcal{D}) = \sup_{x \in \mathcal{C}} \inf_{y \in \mathcal{D}} \|x - y\|_E.$$

Definition 1.2.

A set $\mathcal{A}$ belonging to a Banach space E is called a maximal (global) attractor of a semigroup $\{S_t\}$ ($S_t : E \to E$ for all $t \geq 0$) if this set satisfies the conditions:

1. $\mathcal{A}$ is a compact set in E.
2. $\mathcal{A}$ is the attracting set of the semigroup $\{S_t\}$.
3. $\mathcal{A}$ is strictly invariant:

$$S_t \mathcal{A} = \mathcal{A} \quad \text{for each } t \geq 0.$$

We recall the existence theorem for a maximal attractor (see for example Hale [1], Temam [1], Babin & Vishik [1]).

Theorem 1.1.

Let $\{S_t\}$ be a semigroup of operators acting in a real Banach space E and suppose for each $t \geq 0$ the operator S_t is continuous. If $\{S_t\}$ has a compact attracting set K, then this semigroup has the maximal attractor $\mathcal{A}$, where $\mathcal{A} \subseteq K$.

Definition 1.3.

A set $B_0 \subseteq E$ is called an absorbing set if for any $B \in \mathcal{B}(E)$ there exists such $T = T(B)$ that for $t \geq T$ we have $S_t B \subseteq B_0$.

Since an absorbing set is also an attracting set, a semigroup $\{S_t\}$ of continuous operators having a compact absorbing set has, by theorem 1.1, a maximal attractor $\mathcal{A}$.

Let us note that for a semigroup $\{S_t\}$ having an absorbing set we have proved the existence of the maximal attractor with fewer restrictions on $\{S_t\}$ than in Theorem 1.1 (see Babin & Vishik [1]). The existence of the maximal attractor $\mathcal{A}$ for equation of the type (1), its structure, and estimation of its Hausdorff and fractal dimension have been intensively studied (see again for example Hale [1], Temam [1], Babin & Vishik [1] and bibliographies in these books).

§2 Invariant manifolds

We recall here some facts we need in subsequent sections.

Definition 2.1.
A functional $P : E \to \mathbb{R}$ is said to be a Lyapunov function of a semigroup $\{S_t\}$ if $P(S_t u_0)$ is a decreasing function of t for all $u_0 \in E$, and if u_0 is not an equilibrium point of $\{S_t\}$: $P(S_t u_0) < P(u_0) \ \forall t > 0$.

Definition 2.2.
A point z is called an equilibrium point of $\{S_t\}$ if $S_t z = z \ \forall t \geq 0$.

Definition 2.3.
An equilibrium point z is said to be a hyperbolic point of $\{S_t\}$ if the following conditions hold:

1. The operators S_t ($t \geq 0$) are of class $C^{1+\beta}$, $\beta > 0$, in u in a neighbourhood of the point z.
2. The operators $S'_t(z)$ ($t \geq 0$) (Fréchet derivatives of the operators S_t at the point z) form a linear semigroup on E.
3. The spectrum of the operator $S'_t(z)$ with $t > 0$ does not include points lying on the unit circle $\{\zeta : |\zeta| = 1\}$.
4. The invariant subspace E_+ of the operator $S'_t(z)$, corresponding to the component of the spectrum of $S'_t(z)$ lying outside the unit circle does not depend on t and is finite-dimensional. Its dimension is designated by ind z (the index of the point z).

We use semigroups $\{S'_t(z)\}$ of linear operators $S'_t(z) : E \to E$.

Denote by $\sigma_+(z, \rho)$ the component of the spectrum $\sigma = \sigma(S'_1(z))$ of the operator $S'_1(z) = S'_t(z) \mid_{t=1}$ lying in the domain $\{\zeta \mid |\zeta| > \rho\}$ and by $\sigma_-(z, \rho)$ the component of the spectrum of $S'_1(z)$ lying in the domain $\{\zeta \mid |\zeta| \leq \rho\}$. By $E_+(z, \rho) = E_+$ we denote the invariant subspace of $S'_1(z)$ corresponding to $\sigma_+(z, \rho)$ and by $E_-(z, \rho) = E_-$ the invariant subspace corresponding to $\sigma_-(z, \rho)$. We call ρ a regular spectral radius of $\{S'_t(z)\}$ if

$$\sigma_+(z, \rho) \cap \sigma_-(z, \rho) = \emptyset, \quad E_+(z, \rho) \cap E_-(z, \rho) = \emptyset.$$

If ρ is a regular spectral radius we denote the projection onto E_+ by Π_+, the projection onto E_- by Π_-, Ker $\Pi_+ = E_-$, Ker $\Pi_- = E_+$.

Definition 2.4.

A semigroup $\{S'_t(z)\}$ of linear operators is said to be almost stable if the following conditions hold:
1. The norm of $S'_t(z)$ in $\mathcal{L}(E, E)$ is bounded in E as $t \in [0, T]$ by a constant, depending only on T.
2. $\langle S'_t(z)v, \phi \rangle$ is continuous in $t \geq 0$, $\forall v \in E, \forall \phi \in G^*$ where G^* is a dense set in the conjugate space E^*, $\langle , \rangle$ is the duality between E and E^*.
3. There exists a regular spectral radius $\rho_0 > 0$, $\rho_0 < 1$, such that the set $\sigma_+(z, \rho_0)$ consists of a finite number of points and $E_+(z, \rho_0)$ is finite-dimensional.

Now we formulate the fact we need in theory of linear semigroups (see, for instance, Marsden & McCracken [1]) in the form of the following statement.

Statement 2.1.

Let $\{S'_t(z)\}$ be an almost stable semigroup. If ρ is a regular spectral radius of $\{S'_t(z)\}$ then the subspace $E_+(z, \rho)$ and $E_-(z, \rho)$ are invariant with respect to $S'_t(z)$ for all $t \geq 0$. The restrictions of the operators $S'_t(z)$ on $E_+(z, \rho)$ are invertible operators. There exist norms $\| \cdot \|_+$ and $\| \cdot \|_-$ on E_+ and E_- such that in these norms

$$\| S'_t \mid_{E_-} \|_- \leq (\rho - \varepsilon_-)^t, \tag{2.1}$$

$$\| (S'_t \mid_{E_+})^{-1} \|_+ \leq (\rho + \varepsilon_+)^{-t}, \tag{2.2}$$

where $\varepsilon_-, \varepsilon_+ > 0$, $S'_t = S'_t(z)$. Moreover, the norm $\|u\|_0$ of $u = u_+ + u_- \in E$, $u_+ \in E_+$, $u_- \in E_-$, defined by

$$\|u\|_0 = \max(\|u_+\|_+, \|u_-\|_-), \tag{2.3}$$

is equivalent to the original norm in E:

$$\|u\|_E \leq C\|u\|_0, \quad \|u\|_0 \leq C_1\|u\|_E. \tag{2.4}$$

We consider below such semigroups $\{S_t\}$ as are close to linear in a neighbourhood of any equilibrium point.

Definition 2.5.

A semigroup $\{S_t\}$ is said to be close to linear in a neighbourhood of an equilibrium point z if there exists a neighbourhood $\mathcal{O}_r(z)$ of radius r in the norm $\|\cdot\|_0$ defined by (2.3) of this point such that the semigroup $\{S_t\}$ can be extended beyond this neighbourhood as a semigroup $\{S_t^0\}$ ($S_t^0 u = S_t u$, $S_t u = \mathcal{O}_r(z)$) defined on the whole E. Such a semigroup $\{S_t^0\}$ must have the following properties:

1. Let $u_0 \in \mathcal{O}_r(z)$. Then $S_t u_0 \in \mathcal{O}_r(z)$, $\forall t \in [0, \tau]$ with some $\tau > 0$ if and only if $S_t^0 u_0 \in \mathcal{O}_r(z)$, $\forall t \in [0, \tau]$ and $S_t u_0 = S_t^0 u_0$.

2. The operators S_t^0 are of class $C^{1+\alpha}$ in u on E and

$$\left.\begin{array}{r} S_t^0(z + v) = z + S_t'(z)v + B_t(v), \\ B_t(0) = 0, \quad B_t'(0) = 0. \end{array}\right\} \tag{2.5}$$

3. The Fréchet derivative of the operator B_t satisfies the next condition of proximity to zero over the whole space:

$$\|B_t'(w)\| \leq \mu, \quad \forall w \in E, t \in [0, 1], \tag{2.6}$$

where μ can be chosen arbitrarily small if the radius r of the neighbourhood $\mathcal{O}_r(z)$ is sufficiently small.

4. The derivative $B_t'(w)$ depends on w Hölder continuously with the exponent $\alpha \in (0, 1]$:

$$\begin{array}{l} \|B_t'(w) - B_t'(v)\|/\|w - v\|^\alpha \leq C_0, \\ \forall w, v : 0 < \|w - v\| \leq \delta, \delta > 0, t \in [0, 1], \end{array} \tag{2.7}$$

where $C_0 = C_0(t)$.

5. $\|S_\tau u - z\| \leq C\|u - z\|$, $\quad \forall \tau \in [0, 1], \forall u \in \mathcal{O}_r(z)$.

The following theorem plays an important role in the next section.

Theorem 2.1.

Suppose a semigroup $\{S_t^0\}$ satisfies conditions (1)-(5) and $\{(S_t^0)'(z)\}$ is almost stable. Let ρ be a regular spectral radius of the operator $(S_1^0)'(z)$. Then as μ in (2.6) is sufficiently small there exist manifolds $M_+(z, \rho)$ and $M_-(z, \rho)$ invariant with respect to $\{S_t^0\}$ of class $C^{1+\alpha}$ and tangent at the point z to the affine subspaces $z + E_+(z, \rho)$ and $z + E_-(z, \rho)$ respectively. These manifolds are the graphs of functions

$$y_+ : E_+(z, \rho) \to E_-(z, \rho), \quad g_- : E_-(z, \rho) \to E_+(z, \rho) \tag{2.8}$$

respectively,

$$M_+(z,\rho) = \{u \in E \mid u = z + u_+ + g_+(u_+), u_+ \in E_+(z,\rho)\}, \quad (2.9)$$
$$M_-(z,\rho) = \{u \in E \mid u = z + u_- + g_-(u_-), u_- \in E_-(z,\rho)\}. \quad (2.10)$$

The functions g_+ and g_- are of class $C^{1+\alpha}$,

$$g_+(0) = 0, \quad g_-(0) = 0, \quad g'_+(0) = 0, \quad g'_-(0) = 0, \quad (2.11)$$
$$\|g'_+(u_+)\| \le q, \quad \|g'_-(u_-)\| \le q, \quad \forall u_+ \in E_+, \forall u_- \in E_-, \quad (2.12)$$

where $q = q(\mu)$, $q \to 0$ as in (2.6) $\mu \to 0$. The following operators are well-defined:

$$V_+(u) = \Pi_+(u - z) - g_-(\Pi_+(u - z)), \quad (2.13)$$
$$V_-(u) = \Pi_-(u - z) - g_+(\Pi_-(u - z)). \quad (2.14)$$

These operators have the following properties: there exists $\varepsilon_0 > 0$ such that if $S_k^0 u_0 = S_1^k u_0 \in \mathcal{O}_r(z)$, $k = 0, 1, ..., \ell$, then for $k = 1, ..., \ell$

$$\|V_+(S_1^k u_0)\| \ge (\rho + \varepsilon_0)^k \|V_+(u_0)\|, \quad (2.15)$$
$$\|V_-(S_1^k u_0)\| \le (\rho - \varepsilon_0)^k \|V_-(u_0)\|, \quad (2.16)$$

where $\| \cdot \| = \| \cdot \|_0$ is the same norm as in (2.3).

For the proof of the existence of M_+, M_- see, for instance, in Hirsh, Pugh & Shub [1], Wells [1], Babin & Vishik [4]. The properties of V_- and V_+ are derived from the properties of g_- and g_+ in Babin & Vishik [1], [3] in the case $\rho = 1$. In the case $\rho \neq 1$ these properties are derived in a similar way.

Remark 1.1.

In the case $\rho = 1$ the conditions of Theorem 2.1 can be made less restrictive. The validity of the inequalities (2.6), (2.7) is needed only for $t = 1$, $v, w \in \mathcal{O}_r(z)$.

Let

$$M_+^r(z,\rho) = M_+(z,\rho) \cap \mathcal{O}_r(z), \quad M_-^r(z,\rho) = M_-(z,\rho) \cap \mathcal{O}_r(z). \quad (2.17)$$

By point 1. of Definition 2.5 the manifolds $M_+^r(z,\rho)$ are locally invariant not only with respect to $\{S_t^0\}$ but also with respect to $\{S_t\}$. Let

$$M^+(z,\rho) = \bigcup_{t \ge 0} S_t M_+^r(z,\rho). \quad (2.18)$$

Obviously the set $M^+(z,\rho)$ consists of trajectories $u(t)$, $u(t) = S_t u_0$, $t \ge 0$ belonging to a finitely-parametrized family of trajectories. Such a trajectory $u(t)$ is uniquely defined by a point $u_0 \in M_+^r(z,\rho)$ and by (2.9) $M_+^r(z,\rho)$ is a manifold, its dimension is equal to dim $E_+(z,\rho)$ and a point u_0 is uniquely defined by its projection $\Pi_+ u_0$ on $E_+(z,\rho)$.

II

Local spectral asymptotics

§3 Spectral asymptotics

We shall investigate the asymptotic behaviour of a trajectory $u(t) = S_t u_0$ of the semigroup $\{S_t\}$, $S_t : E \to E$, in a neighbourhood of an equilibrium point z. We suppose that

$$\lim_{t \to +\infty} u(t) = z.$$

Theorem 3.1.
   Let $u(t)$ tend to an equilibrium point z of $\{S_t\}$ if $t \to +\infty$. Let the semigroup $\{S_t\}$ be close to linear semi-group in a neighbourhood of z and $\{S'_t\}$ be almost stable. Let ρ be a regular spectral radius of the operator $S'_1(z)$. Then there exist $T_0 > 0$ and a trajectory $\tilde{u}(t)$ of the semigroup $\{S_t\}$ lying on the local invariant manifold $M^r_+(z, \rho)$ such that

$$\|u(t) - \tilde{u}(t)\| < C' \rho^t \quad \forall t > T_0. \tag{3.1}$$

Remark 3.1.
   Theorem 3.1 is well known in the linear case. Let $u(t)$ be a solution of a linear equation $\partial_t u = -Au$, where the operator A has a complete orthogonal system of eigenvectors $\{e_j \mid j \in \mathbb{N}\}$, with eigenvalues λ_j, where $\lambda_j \to +\infty$ when $j \to \infty$. Then by the Fourier method we have

$$u(t) = \sum_{j=1}^{\infty} c_j \exp(-\lambda_j t) e_j.$$

Obviously for a partial sum of this series

$$\tilde{u}(t) = \sum_{j=1}^{n} c_j \exp(-\lambda_j t) e_j$$

we have an estimate

$$\|u(t) - \tilde{u}(t)\| < C \exp(-\lambda_{n+1} t).$$

This inequality coincides with (3.1) if we put $\rho = \exp(-\lambda_{n+1})$. The $M_+(z, \rho)$ is in this case an invariant space with the basis $e_1, ..., e_n$.

The function $\tilde{u}(t)$ in (3.1) is called a spectral asymptotic of $u(t)$.

Let us note that in the nonlinear case a manifold $M_+(z, \rho)$ is finite-dimensional in view of the conditions of Theorem 3.1, so the trajectories $\tilde{u}(t)$ of the semigroup $\{S(t)\}$ lying on $M_+(z, \rho)$ form a finitely-parametrized family as in the linear case.

Let us present the main ideas of the proof of Theorem 3.1 (for particulars, see Babin & Vishik [4], [1]). For this proof we shall need two lemmas.

Lemma 3.1.
Let non-negative numbers ξ_ℓ ($\ell = 0, 1, ..., n-1$) satisfy the inequalities

$$\xi_{\ell+1} \le p_1 \xi_\ell + p_2^\ell \eta + \zeta \quad (\ell = 0, 1, ..., n-1), \tag{3.2}$$

where $\eta, \zeta > 0$, $p_1 > 0$, $p_2 \ge 0$ and $p_1 \ne p_2$. Then we have the following estimate for ξ_ℓ ($\ell = 1, ..., n$):

$$\xi_\ell \le \xi_0 p_1^\ell + \eta(p_1^\ell - p_2^\ell)(p_1 - p_2)^{-1} + \zeta(1 - p_1^\ell)(1 - p_1)^{-1}. \tag{3.3}$$

Proof   We multiply (3.2) by $p_1^{-\ell-1}$ and sum from 0 to $k-1$. We get

$$p_1^{-k}\xi_k \le \xi_0 + \eta p_1^{-1} \sum_{\ell=0}^{k-1}(p_2 p_1^{-1})^\ell + \zeta \sum_{\ell=0}^{k-1} p_1^{-\ell-1}$$

$$= \xi_0 + \eta p_1^{-1}(1 - (p_2 p_1^{-1})^k)/(1 - p_2 p_1^{-1})$$
$$+ \zeta p_1^{-1}(1 - p_1^{-k})/(1 - p_1^{-1}).$$

So we have established (3.3), by putting $\ell = k$.

Lemma 3.2.

Let the conditions of Theorem 3.1 be satisfied and $\varepsilon > 0$ be sufficiently small. Suppose $S_t u_0 \in \mathcal{O}_\varepsilon(z)$ when $0 \le t \le n$ $(n \in \mathbb{N})$. Then there exists $v_0 \in \mathcal{O}_r(z) \cap M_+^r(z, \rho)$ (where the value of r is the same as in Theorem 2.1) such that

$$v(t) = S_t v_0 \in M_+^r(z, \rho) \text{ for all } t \in [0, n], \Pi_+ v(n) = \Pi_+ u(n) \qquad (3.4)$$

and the inequalities

$$\|\Pi_+(u(k) - v(k))\| \le \rho^k \|V_-(u(0))\| \quad (k = 0, 1, ..., n) \qquad (3.5)$$

$$\|u(t) - v(t)\| \le C\rho^t \|V_-(u(0))\| \quad \forall t \in [0, n+1] \qquad (3.6)$$

hold. The norm $\|\cdot\| = \|\cdot\|_0$ is the same as in (2.3) and C does not depend on $u(0)$ or on n.

Proof   First we prove the inequalities (3.5) and (3.6) if the function $v(k) \in M_+^r(z, \rho)$ for $k = 0, 1, ..., n$. Since $S_\theta u$ satisfies the Lipschitz condition for $0 \le \theta \le 1$ it is enough to prove inequality (3.6) for $t \in \mathbb{N}$. We put $z = 0$ and denote

$$\Pi_\pm v(k) = v_\pm(k), \quad \Pi_\pm u(k) = u_\pm(k).$$

In view of (2.5) we have

$$u(k+1) - v(k+1) = S_1'(z)(u(k) - v(k)) + B_1(u(k)) - B_1(v(k)). \quad (3.7)$$

Let us apply the operator $(S_1')^{-1}\Pi_+$ to both parts of (3.7). Then from (2.6) and (2.2) follow

$$\|(S_1')^{-1}\Pi_+(u(k+1) - v(k+1))\| \ge \|\Pi_+(u(k) - v(k))\|$$
$$- \mu(\rho + \varepsilon_+)^{-1}\|u(k) - v(k)\|.$$

Since $v(k) = v_+(k) + g_+(v_+(k))$ and $u(k) = u_+(k) + u_-(k)$ we have from this inequality

$$\|u_+(k) - v_+(k)\| \le (\rho + \varepsilon_+)^{-1}\|u_+(k+1) - v_+(k+1)\|$$
$$+ \mu(\rho + \varepsilon_+)^{-1}\left[\|u_+(k) - v_+(k)\| + \|u_-(k) - g_+(v_+(k))\|\right]. \qquad (3.8)$$

As from (2.12)

$$\|u_-(k) - g_+(v_+(k))\| \le \|u_-(k) - g_+(u_+(k))\|$$
$$+ \|g_+(u_+(k)) - g_+(v_+(k))\|$$
$$\le \|V_-(u(k))\| + q\|u_+(k) - v_+(k)\|$$

so from (3.8) it follows that

$$\begin{aligned}
\|u_+(k) - v_+(k)\| \leq &(\rho + \varepsilon_+)^{-1}\|u_+(k+1) - v_+(k+1)\| \\
&+ \mu(\rho + \varepsilon_+)^{-1}(1+q)\|u_+(k) - v_+(k)\| \\
&+ \mu(\rho + \varepsilon_+)^{-1}\|V_-(u(k))\|.
\end{aligned} \tag{3.9}$$

Using (2.16), and supposing that μ is sufficiently small (compared with ε_+) we can write

$$\|u_+(k) - v_+(k)\| \leq \rho^{-1}\|u_+(k+1) - v_+(k+1)\| + C\mu(\rho-\varepsilon_0)^k\|V(u(0))\|. \tag{3.10}$$

The inequality (3.10) coincides with the inequality (3.2) if we replace ξ_ℓ by $\|u_+(n-\ell) - v_+(n-\ell)\|$, p_1 by ρ^{-1}, p_2 by $(\rho - \varepsilon_0)^{-1}$, η by

$$C\mu\|V_-(u(0))\|(\rho - \varepsilon_0)^{n-1}$$

and ζ by zero. Since $u_+(n) = v_+(n)$ we have $\xi_0 = 0$ and from the estimate (3.3) (Lemma 3.1) we can conclude that

$$\begin{aligned}
\|u_+(n-\ell) - v_+(n-\ell)\| &\leq C\mu(\rho - \varepsilon_0)^{n-1}\|V_-(u(0))\|\frac{\rho^{-\ell} - (\rho - \varepsilon_0)^{-\ell}}{\rho^{-1} - (\rho - \varepsilon_0)^{-1}} \\
&\leq C_1\mu\rho^{n-\ell}\|V_-(u(0))\|.
\end{aligned}$$

If μ is sufficiently small, i.e. $\mathcal{O}_r(z)$ is sufficiently small, from this inequality follows the estimate (3.5). Now, to prove (3.6), we apply to both sides of the equality (3.7) the operator Π_-. Since $\|\Pi_- S_1' \Pi_-\| \leq \rho - \varepsilon_-$ we get (using (2.6) and (3.5))

$$\begin{aligned}
\|u_-(k+1) - v_-(k+1)\| \leq &(\rho - \varepsilon_-)\|u_-(k) - v_-(k)\| \\
&+ \mu(\|u_-(k) - v_-(k)\| + \|u_+(k) - v_+(k)\|) \\
\leq &(\rho - \varepsilon + \mu)\|u_-(k) - v_-(k)\| + \mu\rho^k\|V_-(u(0))\|.
\end{aligned}$$

This inequality also coincides with (3.2) if we replace ξ_ℓ by

$$\|u_-(\ell) - v_-(\ell)\|,$$

p_1 by $\rho - \varepsilon_- + \mu$, p_2 by ρ, η by $\mu\|V_-(u(0))\|$ and ζ by zero. According to Lemma 3.1 we obtain the corresponding inequality (3.3), and if μ is sufficiently small, we get

$$\|u_-(k) - v_-(k)\| \leq (\rho-\varepsilon_1)^k\|u_-(0) - v_-(0)\| + C_1\mu\|V_-(u(0))\|\rho^k. \tag{3.11}$$

As $v(0) \in M_+$ we have $v_-(0) = g_+(v_+(0))$ and, further,

$$\|u_-(0) - v_-(0)\| \leq \|u_-(0) - g_+(u_+(0))\| + \|g_+(u_+(0)) - g_+(v_+(0))\|$$
$$\leq \|V_-(u(0))\| + q\|u_+(0) - v_+(0)\|.$$

Owing to (3.5) we obtain $\|u_-(0) - v_-(0)\| \leq C_2\|V_-(u(0))\|$ and then from (3.11) follows the estimate

$$\|u_-(k) - v_-(k)\| \leq C_3\rho^k\|V_-(u(0))\|. \tag{3.12}$$

From (3.5) and (3.12) follows (3.6) for integer t.

So we have proved the estimate (3.6) supposing that there exist $v(0), ..., v(n) \in M_+(z, \rho) \cap \mathcal{O}_r(z)$. The existence of $v(k)$ can be proved by induction if we use the reversibility of S_1 in a neighbourhood of zero on M_+ (the reversibility follows from (2.2)). If ε (and therefore $\|V_-(u(0))\|$) is small we see from (3.6) that $\|v(k)\|$ is also small ($k = n, ..., 0$).

From (3.6) for integer t and from the estimate

$$\|S_\theta u - S_\theta v\| \leq C_4\|u - v\|$$

for $\theta \in [0, 1]$ follows (3.6) for all $t \in [0, n+1]$ (with another constant C). The proof of Lemma 3.2 is concluded.

Proof of Theorem 3.1   Since $u(t) \to z$ if $t \to \infty$, for each $\varepsilon > 0$ there exists $T_0(\varepsilon)$ such that $u(t) \in \mathcal{O}_\varepsilon(z)$ for $t > T_0(\varepsilon)$. Let us take a sequence

$$v_n^n = \Pi_+ u(T_0 + n) + g_+(\Pi_+ u(T_0 + n)) \in M_+(z, \rho).$$

Using Lemma 3.2 with the replacement of $u(t)$ by $u(T_0 + t)$ we obtain a sequence $v_n(t)$, $t \in [0, n]$, with $v_n(n) = v_n^n$ satisfying the estimate

$$\|u(T_0 + t) - v_n(t)\| \leq C\rho^t\|V_-(u(T_0))\| \quad \forall t \in [0, n]. \tag{3.13}$$

In particular $\|v_n(0)\| \leq C\|V_-(u(T_0))\| + C_1$ for each n. Thus we can choose from the bounded sequence $\{\Pi_+ v_n(0)\}$ in the finite-dimensional space E_+ such a sub-sequence $v_{+k}(0)$ that

$$\lim_{k \to \infty} v_{+k}(0) = v_{+\infty}.$$

We put $v_\infty = v_{+\infty} + g_+(v_{+\infty})$. Since the function g_+ is continuous we have $v_k = v_{+k}(0) + g_+(v_{+k}(0)) \to v_\infty$ if $k \to \infty$. In view of the uniform continuity of the operators $S_t u$ with respect to u on $t \in [0, T]$ (see point (1) of Definition 2.4 and (2.6)) $S_t v_k \Rightarrow S_t v_\infty$ on $[0, T]$.

Taking in (3.13) $n = k \to \infty$ we obtain

$$\|u(T_0 + t) - S_t v_\infty\| \leq C\rho^t \|V_-(u(T_0))\| \quad \forall t \geq 0.$$

Defining $S_t v_\infty = \tilde{u}(T_0 + t)$ we can write the estimate (3.1). Theorem 3.1 is proved.

§4   Examples of local spectral asymptotics

We shall study here dynamical systems which correspond to differential equations of the form

$$\partial_t u + Au = 0, \tag{4.1}$$

where A is a differential operator, $u = u(x, t)$, $x = (x^1, ..., x^n)$, $x \in \Omega \in E^n$, and u satisfies on $\partial\Omega$ some boundary conditions. In the following examples corresponding spaces and boundary conditions will be clearly pointed out. We discuss here the solutions $u(t) = u(t, x)$ of the equation (4.1) such that $\|u(t) - z\| \to 0$ if $t \to +\infty$, where z is a stationary solution of (4.1), i.e. $Az = 0$. Such a solution $u(t)$ has a spectral asymptotic described in Theorem 3.1. To apply this theorem to the following examples, we replace the equation (4.1) by the equation

$$\partial_t(u - z) = -\left[A(u) - A(z) - A'(z)(u - z)\right]\phi(\|u - z\|/\delta) - A'(z)(u - z), \tag{4.2}$$

where $\phi(\xi) \geq 0$ is a smooth function, $\phi(\xi) = 1$ if $|\xi| \leq 1$, $\phi(\xi) = 0$ if $|\xi| \geq 2$, and δ is sufficiently small. It is obvious that in $\mathcal{O}_\delta(z)$ the equation (4.2) is equivalent to the equation (4.1).

The equation (4.2) is close to a linear equation. Suppose that $\|u\|_E$ is a function in u of the class $C^{1+\beta}$, where $\beta > 0$, for $\|u\|_E \neq 0$. In all examples given below the equation (4.2) generates a semigroup of operators $\{S_t^0\}$, $S_t^0(u(0)) = u(t)$, defined on the whole space E (we could not guarantee for large t the existence of the solutions of the initial equation (4.1) for arbitrary initial condition $u(0) \in E$, so this equation does not necessarily generate a semigroup). It is clear that if $u(t)$ is a solution of (4.1) and $u(t) \to z$ if $t \to \infty$, then $u(t) \in \mathcal{O}_\delta(z)$ for $t \geq \tau = \tau(\delta)$. For these t, equations (4.1) and (4.2) are equivalent and $u(t) = S_{t-\tau}u(\tau) = S_{t-\tau}^0 u(\tau)$. If δ is sufficiently small then in the examples below the semigroup corresponding to (4.2) is close to a linear semigroup.

Using Theorem 3.1 we will state in these examples the existence of spectral asymptotics $\tilde{u}(t)$ for a solution $u(t)$ of (4.1). When $t \geq T$ this asymptotic lies on a finite-dimensional manifold $M_+(z, \rho)$ of the semigroup $\{S_t^0\}$ corresponding to the equation (4.2). Here ρ is a regular spectral radius of the linear semigroup generated by the equation

$$\partial_t v = -A'(z)v. \tag{4.3}$$

By Theorem 3.1 the estimate

$$\|u(t) - \tilde{u}(t)\| \leq C\rho^t \quad \text{when } t \geq T \tag{4.4}$$

holds.

In the following examples ρ is a regular spectral radius of the operator $S_1'(z)$ if the line $\{\xi \mid \text{Re } \xi = \ln \rho\}$ does not intersect the spectrum of the operator $-A'(z)$ acting in E.

We pass on to concrete examples. We will not verify that these examples satisfy the conditions of Theorem 3.1. Similar proofs can be found in our previous works Babin & Vishik [5], [6], [3]. For example the differentiability of the semigroups corresponding to the equations below was established in Babin & Vishik [3] and [1].

Example 4.1. Navier-Stokes system in $\mathbf{R}^2$ or $\mathbf{R}^3$
Let $\Omega \subseteq \mathbf{R}^n$ be a bounded domain with a smooth boundary $\partial\Omega$;

$$V_0 = \{v \mid v \in (C_0^\infty(\Omega))^n, \text{ div } v = 0\};$$

H_s $(0 \leq s \leq 1)$ is the closure of V_0 in the norm of the Sobolev space $(W_2^s(\Omega))^n$. Let us take the projection of the Navier-Stokes system on H_0, i.e.

$$\partial_t u = \nu\Pi\Delta u + \Pi B(u, u) - \Pi f, \tag{4.5}$$

where Π is the operator of orthogonal projection of $(L_2(\Omega))^n$ on H_0, Δ is the Laplace operator,

$$-B(u, v) = \sum_{i=1}^n u_i \partial_i v, \quad f \in (L_2(\Omega))^n.$$

Properties of the Navier-Stokes system can be found, for example, in the works of Ladyzhenskaya [1], Lions [1] and Temam [1]. We write

$H_s = (-\Pi \cdot \Delta)^{-s/2} H_0$ if $s \geq 0$. Let z be a solution of the stationary equation

$$\nu \Pi \Delta z + \Pi B(z, z) = \Pi f. \qquad (4.6)$$

Theorem 4.1.

Let $n = 2$, let $u(t)$ be a solution of equation (4.5) and $u(t) \to z$ in $E = H_1$ if $t \to \infty$. Then for any ρ such that the line Re $\xi = \ln \rho$ does not intersect the spectrum of the operator $-A'(z)$,

$$-A'(z)v = \nu \Pi \Delta v + \Pi B(v, z) + \Pi B(z, v), \qquad (4.7)$$

there exists a spectral asymptotic $\tilde{u}(t)$ on the manifold $M_+(z, \rho)$ satisfying (4.4) with $E = H_1$.

Theorem 4.2.

Let $n = 3$, let $u(t)$ be a solution of (4.5) and for $t \geq T_0$, suppose we have $u(t) \in H_s$ where $3/2 < s < 2$. Further, let $u(t) \to z$ in H_s if $t \to \infty$, where z is a solution of (4.6). Then for any ρ such that the line Re $\xi = \ln \rho$ does not intersect the spectrum of the operator (4.7) there exists a spectral asymptotic $\tilde{u}(t)$ on the manifold $M_+(z, \rho)$ satisfying (4.4) with $E = H_s$.

Let us note that differentiability of the operators $S_t^0 u$ generated by the equation (4.2) is proved for $s > 3/2$ ($n = 3$).

Remark 4.1.

If the viscosity ν in equation (4.5) is sufficiently large then $u(t)$, any solution bounded in H_s ($s > 3/2$), tends to a stationary solution in H_s.

Remark 4.2.

The rate of approach of $u(t)$ to z when $t \to \infty$ was studied by Foias & Saut [1], Foias & Guillopé [1], and Haraux [1], [2]. If $u(t)$ lies on the stable invariant manifolds $M_-(z, \rho)$ of finite-codimension (see §2) then in inequality (4.4) we can replace $\tilde{u}(t)$ by z.

Example 4.2.   Reaction-diffusion system

Let $\Omega \subseteq \mathbf{R}^n$ be a bounded domain. We consider a system

$$\partial_t u = a\Delta u + f(u) - g(x) \qquad (4.8)$$

with Neumann boundary condition

$$\frac{\partial u}{\partial \nu} \Big|_{\partial \Omega} = 0.$$

Here $u = (u^1, ..., u^m)$, $f = (f^1, ..., f^m)$, all $f^i \in C^2$ and the matrix a is symmetric and positive definite. A function f satisfies the condition

$$|\nabla f^i(u) - \nabla f^i(v)| < C(1 + |u|^p)|u - v|^\alpha \quad (i = 1, ..., m) \qquad (4.9)$$

where $\alpha > 0$, $p \geq 0$, $1/(p + \alpha) > (n - 2)/4$, and some other conditions. We suppose also that $g \in (L_2(\Omega))^m$. Let us put

$$H_s = (-\Delta + I)^{-s/2}(L_2(\Omega))^m, \quad E = H_1.$$

In the works of Babin & Vishik [5], [6], [1] are given the conditions of the form (4.9) which enable the equation (4.8) to produce a semigroup $\{S_t\}$, $S_t : E \to E$, and there are studied the properties of this semigroup.

Theorem 4.3.
    Let $u(t)$ be a solution of the equation (4.8) and $u(t) \to z$ in E if $t \to \infty$. Here z is a stationary solution of (4.8). Let the line $\operatorname{Re} \xi = \ln \rho$ not intersect the spectrum of the operator $-A'(z)$, where

$$-A'(z)(v) = a\Delta v + f'(z)v.$$

This operator acts on E. Then there exists a spectral asymptotic $\tilde{u}(t)$ lying on $M_+(z, \rho)$ and satisfying the estimate (4.4).

    The statement of Theorem 4.3 holds in the case of the Dirichlet boundary condition $u|_{\partial\Omega}$ for the equation (4.8).

    Example 4.3.   Quasi-linear parabolic equation of the second order

$$\left. \begin{array}{l} \partial_t u = \sum_{i,j} a_{ij}(x, u, \nabla u)\partial_i\partial_j u + b_0(x, u, \nabla u), \\[2mm] u|_{\partial\Omega} = 0, \quad u|_{t=0} = u_0. \end{array} \right\} \qquad (4.10)$$

It is supposed that the functions $a_{ij}, b_0 \in C^3$ in all variables and

$$\sum_{ij} a_{ij}\xi_i\xi_j \geq \mu_0|\xi|^2, \quad \mu_0 > 0.$$

Then for any regular $\rho > 0$ there exists an asymptotic $\tilde{u}(t)$ on $M_+(z, \rho)$ such that the estimate (4.4) is true for $E = W_p^2(\Omega)$, $p > n$.

III

Global spectral asymptotics

§5 Global spectral asymptotics of trajectories

We shall construct for all $t \geq 0$ asymptotics of trajectories of a semigroup $\{S_t\}$ which is uniform in respect to the bounded set B of initial data $u_0 = u(0)$. We suppose that the semigroup $\{S_t\}$, $S_t : E \to E$, has a finite number of equilibrium points. (Some of them might be unstable.) Any trajectory of such a semigroup tends as $t \to +\infty$ to one of the equilibrium points and has in the neighbourhood of this point a local spectral asymptotic, which was studied in §3. An asymptotic for a trajectory $u(t)$, uniform with respect to initial functions $u_0 = u(0)$, can be different from the individual one studied in §3 because, while $u(t) \to z^m$ if $t \to +\infty$, this trajectory $u(t)$ can, for arbitrarily long times (different for different $u_0 \in B$), stay in neighbourhoods of unstable equilibrium points $z^i \neq z^m$. So, for uniform approximation of $u(t)$, we must use trajectories not only lying on $M_+(z^m, \rho_m)$ but also lying on other manifolds $M_+(z^i, \rho_i)$, $i \neq m$. Therefore the constructed approximation curve has discontinuities (points of discontinuity lie in neighbourhoods of unstable equilibrium points).

Here we describe only a general schema of construction of global spectral asymptotics. For detailed description of these results, see Babin & Vishik [4], [1].

Definition 5.1.
Let $\mathcal{O}_\delta(\mathcal{M})$ be a δ-neighbourhood of the set $\mathcal{M}$ of equilibrium points of $\{S_t\}$ and let $B \subseteq E$ be a bounded set. The number T^0 is said to be a time of arrival (for $\{S_t\}$) from B to $\mathcal{O}_\delta(\mathcal{M})$ if $\forall u_0 \in B$, $\exists t \in [0, T^0]$ such that $S_t u_0 \in \mathcal{O}_\delta(\mathcal{M})$.

Definition 5.2.

A finite-dimensional combined trajectory (f.d.c.t.) is a function $\tilde{u}(t)$ having the following properties: there exist m ($m \leq N$, where N is the number of all equilibrium points) values of t, $t = t_1^0, ..., t_m^0$, $0 = t_0^0 < t_1^0 < ... < t_m^0 < t_{m+1}^0 = \infty$, such that $\forall t \in [t_i^0, t_{i+1}^0)$,

$$\tilde{u}(t) = S_{t-t_i^0}\tilde{u}(t_i^0 + 0), \quad \tilde{u}(t_i^0 + 0) \in M_+(z^i, \rho_i) \equiv M_+(z^i)$$

where $z^i = z_{j(i)}$. As $t \in [t_i^0, t_{i+1}^0)$ the corresponding piece of the curve $\tilde{u}(t)$ lies on $M^+(z^i)$ because

$$\tilde{u}(t_i + 0) \in M_+(z^i) \subseteq M^+(z^i) \equiv M^+.$$

As $M^+(z^i)$ is finite-dimensional and its dimension is $n_j = \text{ind } z_{j(i)}$ there exists an N-parameter family of trajectories of $\{S_t\}$ lying on $M^+(z^i)$. It follows that all f.d.c.t.'s form a finitely-parametrized family of trajectories.

Let us note that in the general case a Lyapunov function has the following property:

$$\text{If } i \neq k \quad P(z_i) \neq P(z_k) \quad (i, k = 1, ..., N). \tag{5.1}$$

Suppose that if $u_1(t)$ and $u_2(t)$ for $t_0 \leq t \leq t_1$ lie in the ball $B_r = \{u \in E \mid \|u\| \leq R\}$ where R is sufficiently large, then there exists $\alpha = \alpha(R)$ such that

$$\|u_1(t) - u_2(t)\| \leq Ce^{\alpha(t-t_0)}\|u_1(t_0) - u_2(t_0)\|, \quad t_0 \leq t \leq t_1. \tag{5.2}$$

By (2.16) if $u(\tau) = S_\tau u_0 \in \mathcal{O}_\epsilon(z_j) \equiv \mathcal{O}(z_j)$ where $t_1 \leq \tau < t$ and $t - \tau$ is an integer then

$$\|V_-(u(t))\| \leq e^{-\gamma_j(t-t_1)}\|V_-(u(t_1))\|, \tag{5.3}$$

$e^{-\gamma_j} = \rho - \delta$, δ is sufficiently small, $\|\cdot\| = \|\cdot\|_0$ is the same norm as in (2.3). This norm is equivalent to the norm in E and it depends on j.

Suppose that if $u_1(t) = S_t u_{10}$ and $u_2(t) = S_t u_{20}$ lie in $\mathcal{O}(z_j)$ for $t_1 \leq t \leq t_2$ then for these t

$$\|u_1(t) - u_2(t)\| \leq Ce^{\alpha_j(t-t_1)}\|u_1(t_1) - u_2(t_1)\|, \quad \alpha_j \geq 0, \tag{5.4}$$

where C and α_j do not depend on u_{10} and u_{20}. Obviously $\alpha_j \leq \alpha$. Under some supplementary conditions we can put $\alpha_j = \ln \rho_j + \delta$, where ρ_j is the radius of a circle containing the spectrum of $S_1'(z_j)$ and $\delta > 0$ is arbitrarily small if the neighbourhood $\mathcal{O}(z_j)$ is sufficiently small.

Theorem 5.1.
Suppose the semigroup $\{S_t\}$ satisfies the following conditions:
(1) *$\{S_t\}$ satisfies conditions (5.2) and (5.4).*
(2) *The set $\mathcal{M}$ of all equilibrium points of $\{S_t\}$ is finite and at every point $z_j \in \mathcal{M}$ the linear subgroup $\{S_t'(z_j)\}$ is almost stable.*
(3) *$\{S_t\}$ has a continuous Lyapunov function on E.*
(4) *$S_t u$ is continuous on (t, u) if $(t, u) \in \mathbf{R}_+ \times E$.*
(5) *For any $\varepsilon > 0$ and any bounded set $B \subseteq E$ the time of arrival from the set $B_1 = \cup_{t \geq 0} S_t B$ to the ε-neighbourhood of the set $\mathcal{M}$ is finite.*
(6) *The semigroup $\{S_t\}$ is close to a linear semigroup in a neighbourhood of any point $z_j \in \mathcal{M}$.*

Let ρ_j $(j = 1, 2, ..., n)$ be regular spectral radii of the operators $S_t'(z_j)$ (see the definition in §2) and let B be a bounded set. Then for each trajectory $u(t) = S_t u_0$, $t \geq 0$, $u_0 \in B$, there exist non-intersecting neighbourhoods $\mathcal{O}(z_j) \subseteq \mathcal{O}_r(z_j)$ of $z_j = z^i$, $j = j(i)$, $i = 1, ..., m$, such that the trajectory $u(t)$ goes consecutively through $\mathcal{O}_\varepsilon(z^i)$ $(j = j(i)$ can depend on a trajectory $u(t))$. There exists also an f.d.c.t. $\tilde{u}(t)$ such that

$$\left.\begin{aligned}
\|u(t) - \tilde{u}(t)\| \leq C_i e^{-\eta^i t}, \quad t_i^- \leq t < t_{i+1}^-, \quad 0 \leq i \leq m, \\
t_0^- = 0, \quad t_j^- \in [t_{j-1}^0, t_j^0), \quad j = 1, ..., m, \\
t_{m+1}^- = +\infty, \quad m = m(u) \leq N,
\end{aligned}\right\} \quad (5.5)$$

t_i^0 are the values of t described in Definition 5.2 and η^i are defined by a recurrence relation,

$$\eta^i = \frac{\gamma^i \eta^{i-1}}{\gamma^i + \eta^{i-1} + \alpha^i} \quad (i = 1, ..., m), \quad (5.5')$$

where η_0 can be taken arbitrary, $C_i = C_i(B, \eta^0)$; and $\gamma^i = \gamma_{j(i)}$, $\alpha^i = \alpha_{j(i)}$ are the same as in (5.3) and (5.4).

Before proving Theorem 5.1 we formulate two lemmas (see Babin & Vishik [1]). Everywhere below we suppose the conditions of Theorem 5.1 hold.

Lemma 5.1.
Let $\mathcal{O} = \mathcal{O}_{\varepsilon_0}(z)$ be an ε_0-neighbourhood of the equilibrium point z in the norm $\| \cdot \|_0$ (see (2.3)). Then there exists $\delta > 0$, $\delta = \delta(\varepsilon_0)$, such that if $u = S_0 u \in \mathcal{O}_\delta(z)$ and $S_\tau u \in \mathcal{O}_\delta(z)$ then the whole segment of trajectory $S_t u$, $0 \le t \le \tau$, belongs to $\mathcal{O}_{\varepsilon_0}(z)$.

Lemma 5.2.
Every point $z_j \in \mathcal{M}$ has an ε-neighbourhood $\mathcal{O}_\varepsilon(z_j)$, where $\varepsilon > 0$ is sufficiently small, such that for any trajectory $u(t)$ the following conditions hold:
(1) $\mathcal{O}_\varepsilon(z_j) \cap \mathcal{O}_\varepsilon(z_i) = \emptyset$ if $i \ne j$.
(2) Let $t_j^- = \inf\{t \mid u(t) \in \mathcal{O}_\varepsilon(z_j)\}$, $t_j^+ = \sup\{t \mid u(t) \in \mathcal{O}_\varepsilon(z_j)\}$, $I_j = [t_j^-, t_j^+]$. Then $I_j \cap I_k = \emptyset$ as $j \ne k$.
(3) $\forall j \le N$, $\exists \varepsilon_j$ such that $P(u(t_j^+)) \le P(z_j) - \varepsilon_j$ where $P(u)$ is the Lyapunov function of the semigroup $\{S_t\}$.

Proof of Theorem 5.1   From Lemmas 5.1 and 5.2 and from condition (5) of Theorem 5.1 it follows that for any trajectory $u(t) = S_t u_0$, $u_0 \in B$, the half-axis $t \ge 0$ contains non-intersecting segments I_i ($i = 1, 2, ..., m$), $m \le N$, and intervals between two neighbouring segments are no longer than some T_1. The construction of the f.d.c.t. $\tilde{u}(t)$ will be done by induction moving along segments I_j and these intervals. Let a trajectory $u(t) = S_t u_0$ go consecutively through neighbourhoods $\mathcal{O}_\varepsilon(z^1), ..., \mathcal{O}_\varepsilon(z^m)$, $z^1 = z_{j(1)}, ..., z^m = z_{j(m)}$, $j(i) \in \{1, 2, ..., N\}$. For $t \in [0, t_1^-]$ we put $\tilde{u}(t) = z^1$. Suppose now that $\tilde{u}(t)$ is constructed for $t \le t_k^-$ and $\forall i \le k-1$ the inequality (5.5) is true. To construct $\tilde{u}(t)$ for $t \in [t_k^-, t_k^+]$ we take two curves $\tilde{u}_k(t)$ and $\tilde{u}_{1k}(t)$, where

$$\tilde{u}_k(t) = S_{t - t_k^-}(\tilde{u}(t_k^-)), \quad t \in [t_k^-, t_k^+],$$

i.e. $\tilde{u}_k(t)$ is the prolongation of the trajectory $\tilde{u}(t)$ constructed for $t \le t_k^-$. As the second trajectory $\tilde{u}_{1k}(t)$ we take the trajectory lying on $M_+(z^k, \rho^k) \equiv M_+$ such that

$$\Pi_+ \tilde{u}_{1k}(t_k^+) = \Pi_+ u(t_k^+).$$

Such a trajectory $\tilde{u}_{1k}(t) \in M_+ \cap \mathcal{O}_r(z^k) \subseteq \mathcal{O}_\varepsilon(z^k)$ for $t \in [t_k^-, t_k^+]$ exists by Lemma 3.2. Now we put

$$\begin{aligned}
\tilde{u}(t) &= \tilde{u}_k(t) \text{ if } t \in [t_k^-, t_k^0], \\
\tilde{u}(t) &= \tilde{u}_{1k}(t) \text{ if } t \in [t_k^0, t_k^+],
\end{aligned} \right\} \tag{5.6}$$

where the value of t_k^0 will be defined later. Let us note that in view of (5.4) and the induction hypothesis we have on $[t_k^-, t_k^+]$

$$
\begin{aligned}
\|u(t) - \tilde{u}_k(t)\| &\le \tilde{C}e^{\alpha_k(t-t_k^-)}\|u(t_k^-) - \tilde{u}_k(t_k^-)\| \\
&\le C \cdot C_{k-1}\exp\left(-\eta^{k-1}t_k^- + \alpha^k(t-t_k^-)\right).
\end{aligned}
\tag{5.7}
$$

From (3.6) where $t = 0$ is replaced by t_k^0 and ρ by $e^{-\gamma^k}$ we have

$$
\|u(t) - \tilde{u}_{1k}(t)\| \le \tilde{C}e^{-\gamma^k(t-t_k^0)}\|V_-(u(t_k^0))\| \le \tilde{C}'e^{-\gamma^k(t-t_k^-)}.
\tag{5.8}
$$

We define the value of t_k^0 in (5.6) so that degrees of exponents in (5.7) and (5.8) are the same:

$$
-\eta^{k-1}t_k^- + \alpha^k(t - t_k^-) = -\gamma^k(t - t_k^-).
\tag{5.9}
$$

Hence

$$
t_k^0 = t_k^- + \eta^{k-1}t_k^-/(\alpha^k + \gamma^k).
$$

If the right-hand side of this equality is $\ge t_k^+$ then we put $t_k^0 = t_k^+$.

Let us prove that with this choice of t_k^0 the function $\tilde{u}(t)$ defined by (5.6) satisfies on $[t_k^-, t_k^+]$ the estimate (5.5) with η^k defined by (5.5').

If $t = t_k^0$ we obtain from (5.7)

$$
\begin{aligned}
\|u(t_k^0) - \tilde{u}(t_k^0)\| &\le C \cdot C_{k-1}\exp\left[-\eta^{k-1}t_k^- + \alpha^k\eta^{k-1}t_k^-/(\alpha^k + \gamma^k)\right] \\
&= C \cdot C_{k-1}\exp\left[-\gamma^k\eta^{k-1}t_k^0/(\gamma^k + \eta^{k-1} + \alpha^k)\right] \\
&= C \cdot C_{k-1}\exp\left[-\eta^k t_k^0\right].
\end{aligned}
\tag{5.10}
$$

We have used here the definition of η^k from (5.5').

Our next step is to prove that the function $\tilde{u}(t)$ defined by (5.6) satisfies on $[t_k^-, t_k^+]$ the inequality (5.5). First we prove this estimate for $t \in [t_k^-, t_k^0]$, namely that

$$
\|u(t) - \tilde{u}(t)\| = \|u(t) - \tilde{u}_k(t)\| \le C \cdot C_{k-1}\exp(-\eta^k t).
\tag{5.11}
$$

For $t = t_k^0$ this inequality is true (see (5.10)). Compare the degrees of exponents in (5.7) and (5.11). Obviously

$$
-\eta^{k-1}t_k^- + \alpha^k(t - t_k^-) \le -\eta^k t
$$

because if $t = t_k^0$ both sides of this inequality are equal (see (5.9)) and for $t < t_k^0$ the left side increases and the right side decreases. So for these values of t, (5.11) is true. Now we take $t_k^+ \geq t \geq t_k^0$ and prove that for these t (5.11) follows from (5.8).

To prove this it is enough to show that for $t \geq t_k^0$

$$-\gamma^k(t - t_k^-) \leq -\eta^k t. \tag{5.12}$$

The inequality (5.12) can be transformed as follows:

$$\gamma^k t_k^- \leq (\gamma^k - \eta^k)t;$$

and as $\eta^k < \gamma^k$ (see (5.5′)) it is enough to prove that

$$\gamma^k t_k^- \leq (\gamma^k - \eta^k)t_k^0.$$

After elementary transformations, using (5.9) and (5.5′), we see that, if $t = t_k^0$, the left side of the inequality is equal to its right side. Hence for $t > t_k^0$ (5.12) is also true. From (5.8) and (5.12) follows (5.5).

Now we define $\tilde{u}(t)$ further on the segment $[t_k^+, t_{k+1}^-]$ as a trajectory of the semigroup $\{S_t\}$:

$$\tilde{u}(t) = S_{t-t_k^+}\tilde{u}(t_k^+), \quad t_k^+ \leq t \leq t_{k+1}^-. \tag{5.13}$$

Since the time of arrival $u(t)$ from $\mathcal{O}_\varepsilon(z^k)$ to $\mathcal{O}_\varepsilon(z^{k+1})$ is bounded we have $t_{k+1}^- - t_k^+ \leq T_1$ and from (5.2) with $t_0 = t_k^+$ we obtain

$$\|u(t) - \tilde{u}(t)\| \leq C \exp\left[\alpha(t - t_k^+)\right] \|u(t_k^+) - \tilde{u}(t_k^+)\|$$
$$\leq C \exp\left[\alpha(t - t_k^+)\right] \cdot \tilde{C}_k \exp(-\eta^k t_k^+) \leq C_k \exp(-\eta^k t) \tag{5.14}$$

where $C_k = C \cdot \tilde{C}_k \exp\left[(\alpha + \eta^k)T_1\right]$. This is the inequality (5.5) for $t \in [t_k^+, t_{k-1}^-]$. So we have extended the approximating function from $t \leq t_k^-$ to $t \leq t_{k+1}^-$ for $k = 1, ..., m-1$.

If $k = m$ the value $t_m^+ = +\infty$. Let $u(t)$ find itself in $\mathcal{O}_\varepsilon(z^m)$ at a moment t_m^-. As $t_m^+ = +\infty$ the function $u(t)$ cannot leave $\mathcal{O}_\varepsilon(z^m)$ for $t > t_m^-$, i.e. $u(t) \to z^m$ if $t \to +\infty$. Detailed study of this case is in §3. That is, we are in the conditions of Theorem 3.1 with $z = z^m$. The function $\tilde{u}(t) = \tilde{u}_{1m}(t)$ for $t \geq t_m^0$ constructed as in Theorem 3.1, and $\tilde{u}(t) = S_{t-t_m^-}u(t_m^-)$ for $t_m^- \leq t < t_m^0$ has the desired properties (see estimate (3.1) where $\rho = e^{-\gamma m}$). The value of t_m^0 is defined from (5.9) with $k = m$. Since $u(t) \in \mathcal{O}_r(z^m)$, $\forall t \geq t_m^-$, we have $u(t) \to z^m$ if $t \to +\infty$ and $u(t) \in M_-(z^m, 1)$.

Remark 5.1.

As follows from (5.5′) the constants η^k ($k = 1, 2, ..., m$) in (5.5) depend on η^0, γ^r, α^r. We note that for all k, $\alpha^k \leq \alpha$ where α is the same as in (5.2). If we put $\gamma^k = \alpha^k \gamma$, where $0 \leq a_1 \leq \alpha^k \leq a_2$, take γ sufficiently large and put $\eta^0 = C_0 \gamma$ then from (5.5′) it follows that $\eta^k = C_k \gamma$ ($k = 1, 2, ..., m$) where $0 < b_1 \leq C_k \leq b_2$ and b_1, b_2 do not depend on k.

Now we will give examples of equations having spectral asymptotics.

Example 5.1.

Let us take a reaction-diffusion system with the Neumann boundary condition in $\Omega \Subset \mathbf{R}^n$

$$\partial_t u = \Delta u - f(x, u) \equiv Au, \quad \frac{\partial u}{\partial \nu}\Big|_{\partial\Omega} = 0, \quad u\,|_{t=0} = u_0, \tag{5.15}$$

where $u = (u^1, ..., u^m)$, $f = (f^1, ..., f^m)$. We suppose that a function $f(x, u)$ satisfies a Lipschitz condition and some conditions on growth of the type

$$|\frac{\partial f}{\partial u^k}| \leq C(1 + |u|)^p, \quad k = 1, ..., m, \quad p = \min(4/n, 2/(n-2)),$$

$$f(x, u) \cdot u \geq \mu_0 |u|^{p_0} - C, \quad p_0 > 2, \quad x \in \overline{\Omega}. \tag{5.16}$$

In complete analogy with Babin & Vishik [3], [8], [6], [1] it can be proved that the equation (5.15) generates a semigroup $\{S_t\}$ in a space $E = H^1$. This semigroup belongs to the class $C^{1+\alpha}$ on u (see Babin & Vishik [1]).

Definition 5.3.

A semigroup $\{S_t\}$, $S_t : E \to E$, is said to satisfy the G-condition if it has a finite number of equilibrium points $\{z_1, ..., z_N\} = \mathcal{M}$ and all z_i are hyperbolic.

For the equation (5.15) the G-condition is generic. If it is not satisfied we can satisfy it by a small change in $f(x, u)$ (see Babin & Vishik [1], [3]). In a neighbourhood of an arbitrary equilibrium point z_j the semigroup $\{S_t\}$ can be extended to a semigroup $\{S_t^0\}$ corresponding to the equation of the form (4.2) with $z = z_j$.

Theorem 5.2.

Let $f(x, u) = \nabla_u F(x, u)$, where $\rho_j \leq 1$ are regular spectral radii of $S_1'(z_j)$ for $j = 1, ..., N$, let $\|u\| \leq R$, and let $\{S_t\}$ satisfy the G-condition. Then for any trajectory $u(t) = S_t u_0$ there exists an f.d.c.t. $\tilde{u}(t) \in \cup M^+(z_j, p_j)$ such that

$$\|u(t) - \tilde{u}(t)\| < C(R)e^{-\eta t}, \quad \eta > 0. \tag{5.17}$$

If all ρ_j are chosen sufficiently small the value of η can be made arbitrarily large.

To prove the theorem it is enough to verify the conditions of Theorem 5.1. Conditions (1)-(4) and (6) easily follow from (5.16) and from the suppositions of Theorem 5.2. We illustrate briefly a proof of condition (5), i.e. that for any $\varepsilon > 0$ and any bounded set $B \subseteq E$ the time of arrival from the set $B_1 = \cup_{t\geq 0} S_t B$ to $\mathcal{O}_\varepsilon(\mathcal{M})$ is finite. From (5.16) it follows that B_1 is bounded and that

$$\int_0^\infty \|\partial_t u(t)\|^2 \, dt \leq C, \quad \|u(t)\|_{L_\infty(0,+\infty;H^1)} \leq C \quad (u(t) = S_t u(0)),$$

$$(5.18)$$

where the constant C does not depend on $u(0) \in B_1$. To prove this we use the global Lyapunov function for (5.15), namely

$$P(u) = \int_\Omega \left(\frac{1}{2} |\nabla u|^2 + F(x,u) \right) \, dx,$$

and the equality

$$\int_0^t \|\partial_t u\|^2 \, dt = P(u(0)) - P(u(t)) \quad \forall t > 0. \qquad (5.18')$$

From (5.18) it follows that on any segment $[0, T]$ there exists a point t such that for $u(t) = S_t u(0)$ we have

$$\|\partial_t u(t)\|^2 \leq C/T, \quad \|u(t)\|_1 \leq C, \qquad (5.19)$$

i.e. at this point t the equation (5.15) is close to a stationary equation. From (5.19) it follows that the time T of arrival from any bounded set $B \subseteq H_1$ to $\mathcal{O}_\varepsilon(\mathcal{M})$ is finite and therefore it is true also for the bounded set $B_1 = \cup_{t\geq 0} S_t B$. Indeed, suppose the contrary. Then for $\varepsilon > 0$ there exists a bounded set B in H_1 such that for all $T^0 > 0$ there exists $u \in B$ such that $\forall t \in [0, T^0]$, $S_t u \notin \mathcal{O}_\varepsilon(\mathcal{M})$. We choose a sequence $T_j^0 \to +\infty$ $(j \to +\infty)$ and for every $T = T_j^0$ such that $t = t_j$ that (5.19) holds. For the sequences $v_j = u_j(t_j)$, $\delta_j = 1/T_j^0 \to 0$ $(j \to +\infty)$

$$\|u_j(t_j)\|_1 \leq C, \quad \|\partial_t u_j(t_j)\| \leq \delta_j, \quad u_j(t_j) \in \mathcal{O}_\varepsilon(\mathcal{M}).$$

We choose a sub-sequence $\{t_k\}$ of $\{t_j\}$ such that $u_k(t_k)$ tends in the weak sense in H_1 to a function u_0 and $\partial_t u_k(t_k) \to 0$ as $k \to +\infty$. Let us put in equation (5.15) $t = t_k$ and look on $\partial_t u_k(t_k)$ as a known function. From the conditions satisfied by $f(x, u)$, it follows that the map $u \to f(x, u)$ is continuous from H_1 in $H = (L_2(\Omega))^m$ and compact from H_1 in $H_{-1} = (H_1(\Omega))'$. Therefore there exists a sub-sequence $\{u_\ell(t_\ell)\}$ of $\{u_k(t_k)\}$ such that

$$f(x, u_\ell(t_\ell)) \to f(x, u_0)$$

in H. Hence and from the equation

$$\Delta u_\ell(t_\ell) - u_\ell(t_\ell) = h_\ell, \quad h_\ell = \partial_t u_\ell(t_\ell) + f(x, u_\ell(t_\ell)) - u_\ell(t_\ell),$$

it follows that $h_\ell \to f(x, u_0) - u_0$ in H_{-1} $(\partial_t u_\ell(t_\ell) \to 0$ in $H_{-1})$. As $(\Delta - I)^{-1}$ is bounded from H_{-1} in H_1 we deduce that $u_\ell(t_\ell) \to u_0$ in the norm H_1 and u_0 is a solution of $\Delta u_0 = f(x, u_0)$, $u_0 \mid_{\partial\Omega} = 0$, i.e. $u_0 \in \mathcal{M}$. As $u_\ell(t_\ell) \to u_0$ in H_1 and $u_0 \in \mathcal{M}$ we have a contradiction with the supposition that $u_j(t_j) \notin \mathcal{O}_\varepsilon(\mathcal{M})$. So we have proved that for every bounded set B in H_1 the time of arrival to $\mathcal{O}_\varepsilon(\mathcal{M})$ is finite.

Everything said above is true also when the Neumann boundary condition is replaced by a Dirichlet condition.

Example 5.2.
Let is take in $\Omega \Subset \mathbf{R}^3$ a hyperbolic equation with dissipation,

$$\partial_t^2 u + \gamma \partial_t u = \Delta u - f(u) - g(x), \quad u\mid_{\partial\Omega} = 0, \quad \gamma > 0 \qquad (5.20)$$

($n = 3$ for simplicity), $u\mid_{t=0} = u_0$, $\partial_t u\mid_{t=0} = p_0$.
We suppose that

$$|f'(u)| \leq C(1 + |u|^2), \quad f'(u) > -C, \quad f(u) \cdot u > -C + |u|^{1+\varepsilon}, \quad (5.21)$$

$\varepsilon > 0$. A space E is the energy space of pairs $(u, p) \in H_1^0(\Omega) \times H(\Omega)$. In this space (5.20) generates the semigroup $\{S_t\}$,

$$S_t(u(0), \partial_t u(0)) = (u(t), \partial_t u(t)),$$

where $u(t)$ is the solution of (5.20) (see Babin & Vishik [3], [8], [1]).

Theorem 5.3.
Let the conditions (5.21) be satisfied and the semigroup $\{S_t\}$ satisfy the G-condition. Let

$$\|(u(0), \partial_t u(0))\| \leq R, \quad (u(t), \partial_t u(t)) = S_t(u(0), \partial_t u(0)).$$

Then there exists an f.d.c.t.

$$(\tilde{u}(t), \partial_t \tilde{u}(t)) \in \bigcup M^+(\hat{z}_j, \rho_j) \quad (1 > \rho_j > \exp(-\gamma/2), \quad j = 1, ..., N),$$

such that the estimate

$$\|(u(t), \partial_t u(t)) - (\tilde{u}(t), \partial_t \tilde{u}(t))\|_E \le C \exp(-\eta t) \qquad (5.22)$$

holds when t is not a point of discontinuity of $\tilde{u}(t)$. Here $C = C(R)$ and η depends only on spectra of all operators

$$\Delta - f'(z_j), \quad z_j \in \mathcal{M} \quad (j = 1, ..., N), \quad \hat{z}_j = (z_j, 0) \quad (p_j = 0).$$

The proof is reduced only to verification of the conditions of Theorem 5.1. Conditions (1)-(4) and (6) follow from (5.21). Condition (5) follows from the existence of a bound for the integral of dissipation

$$\int_0^\infty \left(\|\partial_t u(t)\|^2 + \|\partial_t^2 u(t)\|_{-1}^2 \right) dt < C \qquad (5.23)$$

where C depends only on $\|(u(0), \partial_t u(0))\|_E$. The estimate (5.23) is obtained in analogy with corresponding estimates in Babin & Vishik [6], [1].

Example 5.3.
Let us take a finite-dimensional dynamic system

$$\partial_t u = Au, \quad u \in \mathbf{R}^m.$$

We suppose that for the corresponding semigroup $\{S_t\}$ there exists a bounded absorbing set $B_1 \subseteq \mathbf{R}^m$. We suppose also that $Au \in C^{1+\beta}$, $\beta > 0$, that the set of equilibrium points $\mathcal{M}$ is finite ($\mathcal{M} = \{z_1, ..., z_N\}$, $Az_j = 0$) and that each z_j is hyperbolic. Let there exist a global Lyapunov function (as, for example, when $Au = \nabla_u a(u)$, $a : \mathbf{R}^m \to \mathbf{R}$). Then for any trajectory $u(t) = S_t u(0)$, $t \ge 0$, there exists an f.d.c.t. $\tilde{u}(t) \in \cup_i M^+(z_i, \rho_i)$, where ρ_i are regular spectral radii, and an estimate

$$\|u(t) - \tilde{u}(t)\| \le C \exp(-\eta t), \quad \eta > 0, \quad C = C(\|u(0)\|),$$

holds; $\eta \to \infty$ when $\rho_j \to 0$ $(j = 1, ..., N)$.

IV

Uniform approximation of trajectories of semigroups depending on a parameter

§6 The principal asymptotic term

Let $\{S_t(\lambda) \mid t \geq 0\}$ be a semigroup depending on a parameter λ, $|\lambda| \leq \lambda_0$, $S_t(\lambda) : E \rightarrow E \ \forall t \geq 0$, where E is a Banach space. The trajectories $u(t, \lambda) = S_t(\lambda)u_0$, $t \in \mathbf{R}_+$, of $\{S_t(\lambda)\}$ depend on λ. In this section we construct a global approximation of $u(t, \lambda)$ by a piecewise continuous trajectory $u(t, 0)$ of the limit semigroup $\{S_t(0) \mid t \geq 0\}$. The second, third and subsequent pieces of the trajectory $u(t, 0)$ lie on finite-dimensional manifolds. The estimate

$$||u(t, \lambda) - u(t, 0)|| \leq C|\lambda|^q, \quad q > 0, \quad \forall t \geq 0,$$

$C = C(||u_0||_E)$, holds. Such an approximation $u(t, 0)$ is called a stabilized asymptotic of $u(t, \lambda)$. We shall now give some definitions we need in what follows.

Definition 6.1.
Let $z_1, ..., z_k$ be equilibrium points of the semigroup $\{S_t\}$ and $\mathcal{O}(z_1)$, ..., $\mathcal{O}(z_k)$ the neighbourhoods of these points. $S_t u_0$ is said to pass the neighbourhoods in inverse order if the inclusions $S_t u_0 \in \mathcal{O}(z_i)$, $S_\tau u_0 \in \mathcal{O}(z_j)$ with $\tau \geq t$ imply $j \leq i$.

We suppose the following condition holds.

Condition 6.1.
The semigroup $\{S_t(0)\} = \{S_t\}$ restricted to a set $B_1 \subseteq E$ and the semigroups $\{S_t(\lambda)\}$, $|\lambda| \leq \lambda_1$, restricted to a set $B \subseteq B_1$ have the following properties.

(1) The set $\mathcal{M}$ of all equilibrium points z_i of $\{S_t\}$ is finite,

$$\mathcal{M} = \{z_1, ..., z_N\} \quad \text{and} \quad \mathcal{M} \subset B_1.$$

(2) $\forall \delta > 0$, $\exists T^0 < +\infty$ such that T^0 is a time of arrival for $\{S_t\}$ from B_1 to $\mathcal{O}_\delta(\mathcal{M})$ (see §5).

(3) $\forall u_0 \in B_1$ there exists an enumeration of points of the set $\mathcal{M}$ and there exists $\delta > 0$ (may be dependent on u_0) such that $S_t u_0$ passes $\mathcal{O}_\delta(z_i)$ in reverse order of the enumeration.

(4) Any point $z_j \in \mathcal{M}$ is hyperbolic.

(5) $S_t(\lambda)B \subseteq B_1$, $\forall t \geq 0$, $\forall \lambda \in [-\lambda_1, \lambda_1]$.

(6) There exist constants α and C such that

$$\|S_t(\lambda)u_0 - S_{t-\tau}v_0\| \leq Ce^{\alpha(t-\tau)} \left(|\lambda|^\beta + \|S_\tau(\lambda)u_0 - v_0\| \right) \quad (\beta > 0), \tag{6.1}$$

for any $u_0 \in B$, $v_0 \in B_1$, $\forall t, \tau \geq 0, t \geq \tau$, $\forall \lambda$, $|\lambda| \leq \lambda_1$.

(7) $S_t(\lambda)u_0$ is continuous in t $\forall t \geq 0$, $\forall u_0 \in B$.

Our purpose is to construct under condition 6.1 a global approximation for all $t \geq 0$ of the function $u(t, \lambda) = S_t(\lambda)u_0$ in terms of piecewise continuous trajectories of the limit semigroup $\{S_t\}$.

Definition 6.2.

The family of combined limit trajectories (c.l.t.) corresponding to a family

$$U_\lambda = \{u(\cdot, \lambda) \mid u(t, \lambda) = S_t(\lambda)u_0, u_0 \in B\}, \quad |\lambda| \leq \lambda_1,$$

is a family of trajectories $\tilde{u}(t, \lambda)$ of the semigroup $\{S_t\}$ $(= \{S_t(0)\})$ piecewise continuous in t having the following properties.

(1) The function $\tilde{u}(t, \lambda)$ is continuous in $t \in \mathbf{R}_+$ everywhere excluding the discontinuity points $T_1, ..., T_m$, $T_i = T_i(\tilde{u})$, $T_1 < T_2 < ... < T_m$, $m = m(\tilde{u})$.

(2) The values $\tilde{u}(T_i - 0, \lambda)$, $\tilde{u}(T_i + 0, \lambda)$ at the points of discontinuity T_i lie in a small neighbourhood $\mathcal{O}_r(\mathcal{M})$ of the set $\mathcal{M} = \{z_1, ..., z_N\}$ of all equilibrium points of $\{S_t\}$; $\tilde{u}(T_i + 0, \lambda) \in M_+(z_k)$ for some k.

(3) In any $\mathcal{O}_r(z_k)$ there are values $\tilde{u}(T_i \pm 0)$ at no more than one point of discontinuity T_i. Moreover, if $\tilde{u}(T_i + 0) \in M_+(z_k)$, $k = k(i)$, then $\tilde{u}(t, \lambda) = S_{t-T_i}\tilde{u}(T_i + 0, \lambda)$ $\forall t \in [T_i, T_{i+1})$.

(4) $\tilde{u}(0, \lambda) = u(0, \lambda)$ for some $u(\cdot, \lambda) \in U_\lambda$, $|\lambda| \leq \lambda_1$.

Remark 6.1.
Point (3) of Definition 6.2 implies that on each interval (T_i, T_{i+1}) $(i = 1, ..., m-1)$ the function $\tilde{u}(t, \lambda)$ belongs to a finitely-parametrized family of trajectories of the $\{S_t\} = \{S_t(0)\}$ lying on $M^+(z_k) = \cup_t S_t M_+(z_k)$.

Theorem 6.1.
Suppose Condition 6.1 holds. Suppose the family U_λ corresponds to a set B as in Condition 6.1. Then there exist small numbers $\lambda_0 > 0$, $q > 0$ and a sufficiently large number C such that if $|\lambda| < \lambda_0$ for any $v(\cdot, \lambda) \in U_\lambda$ (that is $v(\cdot, \lambda) = S_t(\lambda)u_0$, $u_0 \in B$) there exists a combined limit trajectory $\tilde{u}(t, \lambda)$ satisfying the inequality

$$\sup_{t \geq 0} \|v(t, \lambda) - \tilde{u}(t, \lambda)\|_E \leq C|\lambda|^q. \tag{6.2}$$

Here q depends on α from (6.1) and also on distances from the spectrum of differentials $S_1'(z_j)$, $z_j \in \mathcal{M}$, to the unit circle. The number C depends on the action of operators S_t in the neighbourhood $\mathcal{O}_\delta(z_j)$, on the value T^0 in Condition 6.1, on constants C and α in (6.1) and on λ_0 and does not depend on λ for $|\lambda| \leq \lambda_0$ or on $u_0 \in B$.

For simplicity we suppose that $\beta = 1$ in (6.1).

Before proving Theorem 6.1 we formulate some lemmas (for detailed proof of them see Babin & Vishik [1]).

Lemma 6.1.
Suppose $\{S_t(\lambda)\}$ satisfies Condition 6.1 on B and $\exists T^0 < +\infty$, where T^0 is the time of arrival for $\{S_t\}$ from $B_1 = \cup_t S_t B$ to $\mathcal{O}_\varepsilon(\mathcal{M})$ $(\mathcal{M} = \mathcal{M}(\{S_t\}))$. Then there exists $\lambda_0 > 0$ such that $\forall \lambda, |\lambda| < \lambda_0$, T^0 is a time of arrival for $\{S_t(\lambda)\}$ from $\cup_t S_t(\lambda)B$ to $\mathcal{O}_{2\varepsilon}(\mathcal{M})$.

Lemma 6.2.
Suppose $\mathcal{M} = \mathcal{M}(\{S_t\})$ is finite, $\mathcal{M} = \{z_1, ..., z_N\}$. Suppose $B_1 \subseteq E$ and $\forall \varepsilon > 0$ a time of arrival for $\{S_t\}$ from B_1 to $\mathcal{O}_\varepsilon(\mathcal{M})$ is finite. Suppose $\exists \varepsilon_0 > 0$ such that if $\varepsilon < \varepsilon_0$ then $\forall u_0 \in B_1$ $S_t u_0$ passes $\mathcal{O}_\varepsilon(z_i)$ in reverse order with respect to the enumeration corresponding to that u_0. Suppose (6.1) is valid on B and B_1 and $B \subseteq E$ satisfies point (5) of Condition 6.1. Then $\exists \lambda_0 > 0$, $\lambda_0 < \lambda_1$ such that $S_t(\lambda)u_0 = u(t, \lambda)$ passes $\mathcal{O}_\varepsilon(z_j)$ in reverse order with respect to the enumeration $\forall \lambda \in [-\lambda_0, \lambda_0]$, $\varepsilon \leq \varepsilon_0/2$, $\forall u_0 \in B$.

Lemma 6.3.
Suppose the conditions of Lemma 6.2 hold, $\mathcal{O}_\varepsilon(\mathcal{M}) \subseteq B$ and suppose, when $\lambda = 0$, $\forall \delta > 0$ there exists $\gamma > 0$ such that for $u(t) = S_t u_0$ we have

$$\text{if } u(t_1), u(t_2) \in \mathcal{O}_\gamma(z_i) \text{ then } u(t) \in O_\delta(z_i) \quad \forall t \in [t_1, t_2]. \tag{6.3}$$

Then $\forall \delta > 0 \; \exists \gamma_1 > 0$ and $\lambda_1 > 0$ such that (6.3) holds with $\gamma = \gamma_1$ for $u(t, \lambda) = S_t(\lambda) u_0$ if $|\lambda| < \lambda_1$.

Lemma 6.4.
Suppose points (4)-(6) of Condition 6.1 hold, $|\lambda| \le \lambda_0$ and λ_0 is small. Let $u(t) = u(t, \lambda) = S_{t-\tau}(\lambda) u(\tau) \in \mathcal{O}(z_i) \; \forall t \in [\tau, T]$ where $\mathcal{O}(z_i)$ is a small neighbourhood of a point $z_i \in \mathcal{M}$, $t < T$. Then for all $t \in [\tau, T]$ such that $t - \tau$ is an integer the following inequality holds:

$$\|V_-(u(t, \lambda))\| \le e^{-\gamma(t-\tau)} \|V_-(u(\tau, \lambda))\| + C_1 |\lambda| \tag{6.4}$$

where C_1 does not depend on $\{u(t)\}$, $\gamma > 0$, $\gamma = -\ln(1 - \varepsilon_0)$ (see (2.16) where $\rho = 1$).

Proof Put $\tau = 0$ for brevity and $t - \tau = t = k$. If a neighbourhood $\mathcal{O}$ is sufficiently small and λ is small, then it follows from (6.1) that if $u(t) \in \mathcal{O}$ then $S_1 u(t) \in \mathcal{O}^1$ where $\mathcal{O}^1$ is a small neighbourhood of the point z_i.

Further we suppose the neighbourhood $\mathcal{O}^i = \mathcal{O}^i(z_i)$ is so small, that in it by formulas (2.14), (2.13) operators V_-, V_+ (corresponding to $S_t = S_t(0)$, that is for $\lambda = 0$) are defined. The inequalities (2.16), (2.15) hold. We write them in the form

$$\|V_-(S_k(u))\| \le e^{-k\gamma} \|V_-(u)\|, \quad \gamma > 0, \tag{6.5}$$

$$\|V_+(S_k(u))\| \ge e^{k\beta} \|V_+(u)\|, \quad \beta > 0. \tag{6.6}$$

Here $\| \cdot \| = \| \cdot \|_0$ is the same norm as in (2.16), (2.15); it depends on z_i. Note that if $u = u(t, \lambda) = S_t(\lambda) u_0 \in \mathcal{O}_1$ then

$$\|V_-(u(t + 1, \lambda))\| = \|V_-(S_1(\lambda) u)\|$$
$$\le \|V_-(S_1 u)\| + \|V_-(u(t + 1, \lambda)) - V_-(S_1 u)\|$$

($S_1 = S_1(0)$). Taking into consideration that V_- satisfies Lipschitz condition on $\mathcal{O}^1$, using (6.1) to estimate $\|u(t + 1, \lambda) - S_1 u\|$ and using (6.5) with $k = 1$ we obtain

$$\|V_-(u(t + 1, \lambda))\| \le e^{-\gamma} \|V_-(u(t, \lambda))\| + C' \|u(t + 1, \lambda) - S_1 u(t, \lambda)\|$$
$$\le e^{-\gamma} \|V_-(u(t, \lambda))\| + C'' |\lambda|. \tag{6.7}$$

This inequality holds for $t = 0, 1, ..., n$ if $n \leq T$. Obviously (6.7) has the form of (3.2) where $\xi_\ell = \|V_-(u(\ell, \lambda))\|$, $p_1 = e^{-\gamma}$, $p_2 = 0$, $\eta = 0$, $\zeta = C''|\lambda|$. By Lemma 3.1 we get the estimate (3.3) with $\ell = t$, $p_1^\ell = e^{-\gamma \ell}$. Estimate (6.4) follows from (3.3) ($\tau = 0$).

Lemma 6.5.
Suppose the conditions of Lemma 6.4 hold. Suppose $u(t) = u(t, \lambda) \in \mathcal{O}^1$, $v(t) \in M_+(z) \cap \mathcal{O}^1$ $\forall t \in [\tau, T]$, $T - \tau$ an integer. Suppose finally $\Pi_+ u(t) = \Pi_+ v(t)$. Then

$$\|v(t) - u(t)\|_E \leq C\|V_-(u(\tau))\| + C|\lambda| \quad \forall t \in [\tau, T]. \tag{6.8}$$

The proof of Lemma 6.5 is quite similar to that of Lemma 3.2. Modifications connected with the parameter λ are the same as in Lemma 6.4.

Lemma 6.6.
Let $u \in \mathcal{O}_{\delta_0}(z_i)$ where δ_0 is sufficiently small. Let $T \geq 0$. Then on $M_+(z_i)$ ($\rho = 1$) there exists a trajectory $v(t)$, $0 \leq t \leq T$, such that $\Pi_+ u(T) = \Pi_+ v(T)$, $v(t) \in M_+(z_i) \cap \mathcal{O}^1(z_i)$ where $\mathcal{O}^1$ is the same as in Lemma 6.4.

Proof of Theorem 6.1   For any point $z_j \in \mathcal{M}$, $j = 1, ..., N$, take a neighbourhood $\mathcal{O}^1(z_j)$ such as in Lemmas 6.4 - 6.6. Let $0 < \delta \leq \delta_0(i)$, $i = 1, ..., N$, where δ_0 is the number from Lemma 6.6. Suppose moreover δ is so small that if $u(\tau, \lambda) \in \mathcal{O}_\delta(z_j)$, $u(T, \lambda) = S_{T-\tau}(\lambda)u(\tau, \lambda) \in \mathcal{O}_\delta(z_j)$ then $u(t, \lambda) \in \mathcal{O}^1(z_j)$ $\forall t \in [\tau, T + 1]$. Note that for $\lambda = 0$ such δ exists by Lemma 5.1 and for $|\lambda| \leq \lambda_0$ where λ_0 is small such a δ exists by Lemma 6.3. We denote by T^0 the number from Lemma 6.1, where $2\varepsilon < \delta$ (we take ε and δ the same for all z_j, $j = 1, ..., N$).

Fix a trajectory $\{u(t, \lambda)\} \in U_\lambda$, $u(t, \lambda) = S_t(\lambda)u_0$, $u_0 \in B$. We shall write $u(t, \lambda) = u(t)$, $\tilde{u}(t, \lambda) = \tilde{u}(t)$ for brevity. The curve $u(t)$ passes through neighbourhoods $\mathcal{O}_\delta(z_j)$ of some points $z_j \in \mathcal{M}$. We enumerate these points in the order $u(t)$ enters these neighbourhoods: $z^1, ..., z^m$, $m \leq N$, $z^i = z_{j(i)}$. It is possible to carry out such an enumeration for small $|\lambda|$ by point (3) of Condition 6.1 and by Lemma 6.2. (Note that the enumeration of z^i in the upper index may depend on u). We denote by t_i^- the first moment t at which $u(t) \in \mathcal{O}_\delta(z^i)$ and by t_i^+ the last moment when $u(t) \in \mathcal{O}_\delta(z^i)$, i.e.

$$t_i^- = \sup\{t \mid \forall \tau < t \quad u(\tau) \notin \mathcal{O}_\delta(z^i)\}$$

(if there is no such τ, then $t_i^- = 0$),

$$t_i^+ = \inf\{t \mid \forall \tau > t \quad u(\tau) \notin \mathcal{O}_\delta(z^i)\}$$

(if there is no such τ, then $t_i^+ = +\infty$). Put $t_0^- = 0$ and put $\tilde{u}(t) = S_t u(0)$ when $t \in [0, t_1^-]$. We construct $\tilde{u}(t)$ for all t using induction with respect to i. Suppose $\tilde{u}(t)$ is constructed when $0 \le t \le t_i^-$ and for these values of t the following inequalities holds:

$$\|u(t) - \tilde{u}(t)\|_E \le C_{i-1}|\lambda|^{q_{i-1}}. \tag{6.9}$$

We proceed to construct $\tilde{u}(t)$ for $t \in [t_i^-, t_i^+]$ and if $t_i^+ \ne +\infty$ for $t \in [t_i^-, t_{i+1}^-]$. To simplify notation let $z^i = 0$, $\mathcal{O}_\delta(z^i) = \mathcal{O}_\delta$, $M_+(z^i) = M_+$, $t_i^- = 0$, $t_i^+ = t_0^+$, $t_{i+1}^- = t_1^-$, $T_i = T$.

Consider two curves $\tilde{u}_1(t)$ and $\tilde{u}_2(t)$. Put $\tilde{u}_1(t) = S_t\tilde{u}(0)$ and put $\tilde{u}_2(t) = v(t)$ where $v(t) \in M_+$ as a trajectory of $S_t = S_t(0)$ such that $\Pi_+ v(t_0^+ + \theta) = \Pi_+ u(t_0^+ + \theta)$, $0 \le \theta < 1$, $t_0^+ + \theta$ an integer. Such a trajectory $v(t) \in M_+ \cap \mathcal{O}^1$ exists by Lemma 6.6 when $t \in [0, t_0^+]$. Put

$$\tilde{u}(t) = \tilde{u}_1(t) \text{ when } t \in [0, T] \tag{6.10}$$

$$\tilde{u}(t) = \tilde{u}_2(t) \text{ when } t > T. \tag{6.11}$$

We shall give the value of T later in explicit form. Note that $T \in [0, t_0^+]$ and either T is an integer or $T = t_0^+$. We choose T such that the following inequality holds:

$$\|u(t) - \tilde{u}(t)\|_E \le C_i|\lambda|^{q_i} \tag{6.12}$$

for $t \in [t_i^-, t_i^+] \equiv [0, t_0^+]$. Note that by (6.1) with $\tau = 0$ and by (6.9) with $t = 0$ we have

$$\|u(t) - \tilde{u}(t)\|_E \le Ce^{\alpha t}(|\lambda| + C_{i-1}|\lambda|^{q_{i-1}}) \quad \forall t \in [0, T]. \tag{6.13}$$

In view of (6.8)

$$\|u(t) - \ddot{u}(t)\|_E \le C\|V_-(u(T))\|_E + C|\lambda| \quad \forall t \in [T, t_0^+].$$

Estimating $\|V_-(u(T))\|$ by (6.4) with $\tau = 0$ we obtain

$$\|u(t) - \tilde{u}(t)\|_E \le C'e^{-\gamma^i T} + C'|\lambda| \quad \forall t \in [T, t_0^+] \tag{6.14}$$

where $\gamma^i = \gamma(z^i) = \gamma(z_{j(i)})$. Take as T the integer part of the number σ, $T = [\sigma]$, where σ is defined by the equality

$$e^{\alpha\sigma}|\lambda|^{q_{i-1}} = e^{-\gamma^i\sigma}, \quad \sigma = -q_{i-1}\ln |\lambda|/(\alpha + \gamma^i). \qquad (6.15)$$

From (6.13) and (6.15) since $T \le \sigma$ we obtain

$$\|u(t) - \tilde{u}(t)\|_E \le C|\lambda|^{-q_{i-1}\alpha/(\alpha+\gamma^i)} (|\lambda| + C_{i-1}|\lambda|^{q_{i-1}}).$$

Hence the inequality (6.12) with

$$q_i = q_{i-1}\gamma^i/(\alpha + \gamma^i) \qquad (6.16)$$

holds $\forall t \in [0, T]$. Since $T \ge \sigma - 1$, using (6.14), we obtain

$$\|u(t) - \tilde{u}(t)\|_E \le e^{\gamma^i} C'|\lambda|^{\gamma^i q_{i-1}/(\alpha+\gamma^i)} + C'|\lambda|, \quad \forall t \in [T, t_0^+]. \qquad (6.17)$$

Therefore (6.12), with q_i defined by (6.16), holds $\forall t \in [T, t_0^+]$. In the case $[\sigma] \ge t_0^+$ we put $T = t_0^+$ in (6.10).

Now consider separately the case $t_i^+ = +\infty$, that is $i = m$. This case can be considered in complete analogy to the last step in the construction of spectral asymptotics at a point z^m in the proof of Theorem 3.1. We construct $\tilde{u}(t)$ for $t \ge T_m$ as a limit of trajectories $\tilde{u}_k(t)$ $(k = 1, 2, ...)$ of the semigroup $\{S_t\}$ lying on $M_+(z^m)$; these trajectories satisfy the condition

$$\Pi_+ u(t_k) = \Pi_+ \tilde{u}_k(t_k), \quad t_k \to +\infty.$$

We use the fact that the constants in inequality (6.17) do not depend on t_0^+, i.e. on $t_k = t_0^+$ if $\tilde{u} = \tilde{u}_k$.

Now we define the trajectory $\tilde{u}(t)$ for $t \in [t_0^+, t_1^-]$ $(= [t_i^+, t_{i+1}^-])$ in the case $t_0^+ < +\infty$. Put $\tilde{u}(t) = S_{t-t_0^+}(\tilde{u}(t_0^+ - 0))$ for $t \in [t_i^+, t_{i+1}^-]$. Using (6.12) with $t = t_0^+$ and (6.1) with $\tau = t_0^+$ and using also that $t_1^- - t_0^+ \le T^0$ we obtain, from Lemma 6.1 and point (2) of Condition 6.1, that the inequality (6.12) holds for $0 \le t \le t_1^-$ $(= t_{i+1}^-)$. Therefore (6.12) is true for $i = 1, ..., m$ and Theorem 6.1 is proved.

Remark 6.2.

If the equilibrium point z_i is stable, that is ind $z_i = 0$, then it can coincide only with z^m since $u(t, \lambda) \to z_i(\lambda)$ if $t \to +\infty$ and $z_i(\lambda) \to z^m = z_i$ if $\lambda \to 0$.

Remark 6.3.

It follows from (6.16) that in (6.2)

$$q = q_m = \min\left(\frac{\gamma^{j(1)}}{\alpha + \gamma^{j(1)}} \times \cdots \times \frac{\gamma^{j(m)}}{\alpha + \gamma^{j(m)}}\right) \qquad (6.18)$$

where the minimum is taken over integer-valued functions $j = j(i)$ corresponding to enumerations $z^i = z_{j(i)}$.

Lemma 6.7.

Suppose that for any $u_0 \in E$ the set $\{u(t) = S_t u_0, \quad t \geq 1\}$ is precompact, $\{S_t\}$ admits a continuous Lyapunov function P on E and $S_t u$ is continuous with respect to $(t, u) \in \mathbf{R}_+ \times E$. Suppose also that the set $\mathcal{M}$ of all equilibrium points of $\{S_t\}$ is finite and all the points $z_j \in \mathcal{M}$ are hyperbolic, and the points z_j are enumerated in such a way that

$$P(z_1) \leq \dots \leq P(z_N).$$

Let the neighbourhoods $\mathcal{O}(z_j)$ of the points z_j be sufficiently small. Then $\forall u_0 \in E$ the trajectory $S_t u_0$ passes $\mathcal{O}(z_j)$ in reverse order of the enumeration.

Proof   Let all $\mathcal{O}(z_j)$ be so small that they are inside the neighbourhoods from Lemma 5.2. Moreover, $\mathcal{O}(z_j)$ are so small that $|P(u) - P(z_j)| < \varepsilon_0/2$, if $u \in \mathcal{O}(z_j)$, where ε_0 is the minimum of all numbers ε_j from point (3) of Lemma 5.2. Let $u(t_1) \in \mathcal{O}(z_i)$, $u(t_2) \in \mathcal{O}(z_j)$, $t_2 > t_1$, $i \neq j$. Then, by Lemma 5.2, since $t_2 > t_1$ and $u(t_2) \notin \mathcal{O}(z_i)$ we have

$$P(u(t_2)) \leq P(z_i) - \varepsilon_i \leq P(z_i) - \varepsilon_0.$$

Since $u(t_2) \in \mathcal{O}(z_j)$, $|P(u(t_2)) - P(z_j)| \leq \varepsilon_0/2$. Therefore $P(z_j) \leq P(u(t_2)) + \varepsilon_0/2 \leq P(z_i) - \varepsilon_0/2$. Therefore, by (6.19), $j < i$ and $u(t)$ passes $\mathcal{O}(z_j)$ in reverse order to the enumeration.

Now we give a generalization of Theorem 6.1 which can be used for investigation of evolutionary equations singularly depending on a parameter λ, for example a semilinear hyperbolic equation having the second time derivative with coefficient $\lambda > 0$ (see Example 7.3 in §7).

We denote by U_λ a set of functions $u = v(t, \lambda)$, $t \in [0, +\infty)$ depending on a parameter λ. Let B_λ denote the union of all values (in E) of $v(t, \lambda)$, $B_\lambda = \{u \mid u = v(t, \lambda), v \in U_\lambda, t \in \mathbf{R}_+\}$. (In the examples below U_λ is a set of solutions. For example U_λ is the set of all trajectories $S_t(\lambda)u_0$

with $||u_0|| \leq R$. A more complicated example with singular dependence on a parameter is Example 7.3 in §7.) We consider also a semigroup $\{S_t\}$, $S_t : E \to E$ (it corresponds to $\lambda = 0$).

Instead of Condition 6.1 we introduce Condition (6.1′) with U_λ and $\{S_t\}$. Points (1′) - (4′) of Condition (6.1′) coincide with points (1) - (4) of Condition 6.1. Instead of conditions (5) - (7) we introduce the following conditions:

(5′) $B_\lambda \subseteq B_1$ $\forall \lambda, |\lambda| \leq \lambda_1$ (consequently $\cup_{|\lambda| \leq \lambda_1} B_\lambda \subseteq B_1$).

(6′) *There exist constants α and C such that $\forall v_0 \in B_1$ and $\forall u \in U_\lambda$,*

$$||u(t,\lambda) - S_{t-\tau}v_0|| \leq C e^{\alpha(t-\tau)}(|\lambda|^\beta + ||u(t,\lambda) - v_0||), \qquad (6.20)$$
$$t \geq \tau, \quad |\lambda| \leq \lambda_1.$$

(7′) *Functions $v(t,\lambda)$, $(v(\cdot,\lambda) \in U_\lambda, |\lambda| \leq \lambda_1)$ are continuous in t from $\mathbf{R}_+$ to E.*

The following analogue to Theorem 6.1 is true.

Theorem 6.1′.

Suppose Condition 6.1′ holds. Then there exist small numbers $\lambda_0 > 0$ and $q > 0$ and a sufficiently large C such that for any $u(\cdot,\lambda) \in U_\lambda$ there exists a c.l.t. $\tilde{u}(t,\lambda)$ satisfying

$$\sup_{t \geq 0} ||u(t,\lambda) - \tilde{u}(t,\lambda)||_E \leq C_0 |\lambda|^q \qquad (6.21)$$

where the exponent q depends on α in (6.20) and on the spectrum of differentials $S_1'(z_j)$. The constant C depends on $\{S_t\}$, B_1, C and α in (6.20).

Proof of this theorem is quite similar to the proof of Theorem 6.1. One has only to change $S_t(\lambda)u_0$ everywhere to $u(t,\lambda)$ and use Condition (6.1′) instead of Condition (6.1).

§7 Stabilised asymptotics of solutions of reaction-diffusion systems, hyperbolic and parabolic equations

We consider some systems of partial differential equations depending on a parameter λ. For solutions $u(t,\lambda)$ of these equations we construct stabilized asymptotics.

For brevity we give only the formulations of the corresponding theorems and the ideas of proofs. More detailed proofs one can find in works of Babin & Vishik [3], [8], [1].

Example 7.1.
Consider a system of reaction-diffusion type, containing a parameter:

$$\partial_t u = \Delta u - f(x, u, \lambda) - g(x) \equiv A(u, \lambda), \quad x \in \Omega \subseteq \mathbf{R}^n, \qquad (7.1)$$

where, as in §4, $u = (u^1, ..., u^m)$, $f = (f^1, ..., f^m)$. Let the boundary condition be the Neumann condition

$$\frac{\partial u}{\partial \nu}\big|_{\partial \Omega} = 0.$$

Suppose that a condition of the form (5.16) is satisfied uniformly in λ if $|\lambda| < \lambda_0$ and a Lipschitz condition in λ is satisfied also. As in the works of Babin & Vishik [3], [9], [1], it can be proved that a semigroup $\{S_t(\lambda)\}$, $|\lambda| \le \lambda_0$, acting in the space $E = H^1$, belongs to the class $C^{1+\alpha}$, $0 < \alpha \le 1$.

Theorem 7.1.
Suppose for $\lambda = 0$ $f(x, u, 0) = (f^1, ..., f^m) = \nabla_u F$, i.e. $f^i = \partial F / \partial u^i$, where $F : \mathbf{R}_u^m \to \mathbf{R}$ is of class $C^{2+\alpha}$ in u. Let $\{S_t(0)\}$ satisfy the G-condition. Then for any set B bounded in E there exists $\lambda_0 > 0$ such that if $|\lambda| \le \lambda_0$ and $u_0(\lambda) = u(t, \lambda)|_{t=0} \in B$ then there exists for $u(\lambda) = S_t(\lambda)u_0(\lambda)$ a c.l.t. $\tilde{u}(t, \lambda)$ such that

$$\sup_{t \ge 0} \|u(t, \lambda) - \tilde{u}(t, \lambda)\|_E \le C|\lambda|^q, \quad q > 0. \qquad (7.2)$$

Here $C = C(\|u_0(\lambda)\|)$, $\tilde{u}(t, \lambda)$ has all the properties described in Theorem 6.1.

To prove Theorem 7.1 it is enough to verify the conditions of Theorem 6.1. As U_λ we take the set $\{S_t(\lambda)u \mid u \in B, |\lambda| \le \lambda_0\}$. The finiteness of the time of arrival for $\{S_t(0)\}$ from B to $\mathcal{O}_\delta(\mathcal{M})$ was proved in Example 5.1. The other points of Condition 6.1 can be proved in a similar way. Let us pay attention to the proof of (6.1). Let us subtract from the equation (7.1) for $u = S_t(\lambda)u_0$ the equation for $S_t(0)v_0 = v$, i.e. equation (7.1) with $\lambda = 0$. We obtain

$$\partial_t(u - v) = \Delta(u - v) - [(f(u, \lambda) - f(u, 0)) + (f(u, 0) - f(v, 0))],$$
$$f(u, \lambda) = f(x, u, \lambda).$$

The scalar product of both sides of this equality and the function $-\Delta(u - v)$ after elementary transformations gives the inequality

$$\partial_t \|u - v\|_1^2 \le |\lambda|^2 C + C_1 \|u - v\|_1^2.$$

Hence follows (6.1).

Example 7.2.
Consider in $\Omega \in \mathbf{R}^3$ the hyperbolic equation with dissipation depending on a parameter λ:

$$\partial_t^2 u + \gamma \partial_t u = \Delta u - f(u, \lambda) - g, \quad u|_{\partial\Omega} = 0. \qquad (7.3)$$

Suppose that $f(u, \lambda)$ satisfies conditions (5.21) uniformly in λ if $|\lambda| \le \lambda_0$ and suppose that the Lipschitz condition in λ,

$$|f_u'(u, \lambda_1) - f_u'(u, \lambda_2)| \le C|\lambda_1 - \lambda_2|(1 + |u|^3), \qquad (7.4)$$

is also satisfied.

Theorem 7.2.
Suppose $\{S_t(0)\}$ satisfies the G-condition. Then for any bounded $B \subseteq E$, where $E = H_0^1 \times H$, there exist $C > 0, q > 0$, $\lambda_0 > 0$ such that for $|\lambda| \le \lambda_0$ and $(u(0, \lambda), \partial_t u(0, \lambda)) \in B$ there exists a c.l.t. $(\tilde{u}(t, \lambda), \partial_t \tilde{u}(t, \lambda))$ corresponding to (7.3) with $\lambda = 0$ such that the estimate

$$\|(u(t, \lambda), \partial_t u(t, \lambda)) - (\tilde{u}(t, \lambda), \partial_t \tilde{u}(t, \lambda))\|_E \le C|\lambda|^q, \quad \forall t \ge 0, \qquad (7.5)$$

is true for all t with the exception of points of discontinuity in t of the function $(\tilde{u}(t, \lambda), \partial_t \tilde{(t, \lambda)})$.

The proof of the theorem reduces to verifying Condition 6.1. We describe briefly the proof of point (6) of this condition. As in Example 7.1 we subtract the equation (7.3) for v with $\lambda = 0$ from the equation (7.3) for u with $\lambda \ne 0$. After this we take the scalar product of the equality obtained on $\partial_t(u - v)$ and after elementary transformations obtain the estimate

$$\begin{aligned}
\|(u(\tau) - v(\tau), \partial_t u(\tau) - \partial_t v(\tau))\|_E^2 \le &\|(u(0) - v(0), \partial_t u(0) - \partial_t v(0))\|_E^2 \\
&+ C_1|\lambda|^2 \\
&+ C \int_0^\tau \|(u - v, \partial_t u - \partial_t v)\|_E^2 \, dt.
\end{aligned}$$

From this, using the Gronwall inequality we obtain (6.1).

Example 7.3.

Consider the hyperbolic equation with dissipation depending on a parameter λ singularly, $\lambda \in (0, \lambda_0]$:

$$\lambda \partial_t^2 u + \partial_t u = \Delta u - f(u) - g(x), \quad x \in \Omega, \quad u\,|_{\partial\Omega} = 0. \qquad (7.6)$$

The function $f(u)$ satisfies the same conditions as in Example 5.2 (see (5.21)). The semigroup $\{S_t(\lambda)\}$ corresponding to the equation (7.6) acts in the energy space $E = H_1 \times H$ where H_s is a scale of spaces generated by $-\Delta u$, $u|_{\partial\Omega} = 0$. As $\lambda = 0$ the problem (7.6) degenerates into the parabolic problem

$$\partial_t u = \Delta u - f(u) - g(x), \quad u|_{\partial\Omega} = 0. \qquad (7.7)$$

The semigroup $\{S_t(0)\} = \{S_t\}$, corresponding to (7.7), acts in the space H_1, $S_t : u(0) \to u(t)$. (So the limit of $S_t(\lambda)$ as $\lambda \to 0$ acts in a space different from the space in which $S_t(\lambda)$ acts.) Therefore instead of studying $\{S_t(\lambda)\}$ with $\lambda \neq 0$ we will study a family of sets $U_\lambda = U_\lambda(M_2)$ of solutions $u(t) = u_\lambda(t)$ of the equation (7.6) satisfying the condition

$$\|u_\lambda(0)\|_2^2 + \|\partial_t u_\lambda(0)\|^2 + \lambda \|\partial_t u_\lambda(0)\|_1^2 \leq M_2. \qquad (7.8)$$

For such sets U_λ and for the semigroup $\{S_t\}$ all points of Condition 6.1' are true and hence Theorem 6.1' can be applied. We formulate the basic lemmas that must be used for verifying Condition 6.1'.

Lemma 7.1.

Suppose $u(\cdot) = u_\lambda(\cdot) \in U_\lambda(M_2)$. Then there exist constants $\alpha > 0$, $C > 0$ and $K_1, K_2 > 0$ such that $\forall t \geq 0$ the estimates

$$\lambda \|\partial_t u(t)\|_1^2 + \|u(t)\|_2^2 + \int_0^t \|\nabla \partial_t u(\tau)\|^2 \, d\tau \leq C e^{\alpha t},$$

$$\|\partial_t u(t)\| \leq K_1 e^{\alpha t}, \quad \lambda \int_0^t \|\partial_t^2 u(\tau)\|^2 \, d\tau \leq K_2 e^{\alpha t}$$

are true. Constants α, C, K_1, K_2 depend only on M_2.

Lemma 7.2.

Suppose $u(\cdot) = u_\lambda(\cdot) \in U_\lambda(M_2)$ and u_0 is a solution of (7.7),

$$\|u_0(t)\|_1 \leq C_0 \quad \forall t \geq 0.$$

Then the inequality

$$\|u(t) - u(t_0)\|_1 \leq C(\lambda^{1/2} + \|u(t_0) - u_0(t_0)\|_1)e^{\alpha_1(t-t_0)} \quad (7.9)$$

if $t \geq t_0 > 0$, is true, where $C = C(M_2)$, $\alpha_1 = \alpha_1(M_2, C_0)$.

Let us note that the semigroup $\{S_t\}$ corresponding to (7.7) admits a Lyapunov function

$$P(u_0) = \int_\Omega \left(\frac{1}{2} \|\nabla u_0\|^2 + F(u_0) + gu_0 \right) dx, \quad F'(\xi) = f(\xi),$$

and $P(u)$ is continuous in H_1. Equilibrium points $z(x)$ of the equation (7.7) are defined from the equation

$$A_0(z) \equiv \Delta z - f(z) = g, \quad z|_{\partial\Omega} = 0. \quad (7.10)$$

If g is a regular value of the operator A_0, $A_0 : H_2 \to H$, then this equation has a finite number of solutions $(z_1, ..., z_N) = \mathcal{M}$ and in every point $z_i \in \mathcal{M}$ the differential $A_0'(z_i)v = \Delta v - f'(z_i)v$, $v|_{\partial\Omega} = 0$, is not degenerate. Therefore all the points z_i are hyperbolic, that is $\{S_t\}$ satisfies the G-condition (see Definition 5.3).

Theorem 7.3.
Suppose $\{S_t\}$ satisfies the G-condition and $f(u)$ satisfies conditions (5.21). Suppose $U_\lambda = U_\lambda(M_2)$ is the set of solutions $u(t) = u_\lambda(t)$ of the equation (7.6) satisfying (7.8). Then for any $u_\lambda(\cdot) \in U_\lambda(M_2)$ there exists a combined limit trajectory $\tilde{u}(t, \lambda) = \tilde{u}(t)$ of the limit equation (7.7) such that for sufficiently small λ_0 and $0 < \lambda \leq \lambda_0$

$$\sup_{t \geq 0} \|u_\lambda(t) - \tilde{u}(t)\|_1 \leq C|\lambda|^q, \quad q > 0, \quad C = C(M_2). \quad (7.11)$$

If $t \geq \lambda^{1-\delta}$, $1 > \delta > 0$, and t is not a discontinuity point of $\tilde{u}(t)$ the following estimate holds:

$$\|\partial_t u_\lambda(t) - \partial_t \tilde{u}(t)\|_{-1} \leq C_1|\lambda|^q \quad (7.12)$$

where $C_1 = C_1(M_2)$. The degree q in (7.11) and (7.12) depends on α_1 in (7.9) and also on distances to zero from the spectrum of the derivative $A'(z_i)$ at any point $z_i \in \mathcal{M}$ and on the number N of these points.

The proof of Theorem 7.3 reduces to verifying the conditions of Theorem 6.1' (see Babin & Vishik [1], [4]).

If $g(x)$ and $(u_\lambda(0), \partial_t u_\lambda(0))$ are subjected to additional smoothness conditions the statement of Theorem 7.1 can be maintained.

Remark 7.1.
The dependence on a parameter of attractors of the equation (7.6) has been studied in many works. (See, for example, Hale & Raugel [1], Mora & Sola Morales [1]).

Example 7.4.
Consider a parabolic, singularly perturbed equation

$$\partial_t u = -\lambda \Delta^2 u + \Delta u - f(u) - g, \quad u|_{\partial\Omega} = 0, \quad \Delta u|_{\partial\Omega} = 0, \quad \lambda > 0. \quad (7.13)$$

Suppose $f(u)$ satisfies the usual conditions: $f(u)u \geq -C + \delta^2 u^2$ where δ is sufficiently small, $f'(u) \geq -C_1$ $\forall u \in \mathbf{R}$ and some condition on growth, for example

$$|f(u)| \leq C(1 + |u|^{p-1}), \quad L_p(\Omega) \supseteq H^1(\Omega).$$

We prove the following estimate of type (6.1):

$$\|u_\lambda(t) - u_0(t)\| \leq C(\lambda^{1/2} + \|u_\lambda(\theta) - u_0(\theta)\|)e^{\alpha(t-\theta)} \quad (7.14)$$

where $\forall t \geq \theta$, $\alpha > 0$, $\theta \geq \delta > 0$, $u_\lambda(\cdot)$ is the solution of (7.13) and $u_0(\cdot)$ is the solution of the limit problem for (7.13) as $\lambda = 0$:

$$\partial_t u_0 = \Delta u_0 - f(u_0) - g, \quad u_0|_{\partial\Omega} = 0. \quad (7.15)$$

To prove the inequality (7.14) we first prove an *a priori* estimate for solutions $u(t)$ of (7.13), namely

$$\lambda \int_\theta^t \|u(\tau)\|_3^2 \, d\tau \leq M(|t - \theta| + 1) \quad (7.16)$$

where $\theta \geq \delta > 0$ and M does not depend on $t - \theta$.

Taking the scalar product of both sides of (7.13) with $u(t)$ we can easily prove that

$$\|u(t)\|^2 \leq \|u(\theta)\|^2 e^{-\lambda_1(t-\theta)} + C \quad (7.17)$$

where λ_1 and C do not depend on t, θ or λ. One can prove that, if $t > 0$, the semigroup $\{S_t(\lambda)\}$, corresponding to (7.13), is (H, H_1)-bounded. To prove this we take the scalar product of both sides of (7.13) with $-t\Delta u$ and integrate over t. We obtain an estimate

$$\|\nabla u(t)\|^2 \le C_1(\|u(t)\|^2 + 1) \quad \forall t, \ \delta \le t \le \theta + 1, \quad 0 < \delta < 1. \quad (7.18)$$

If we integrate (over t) the scalar product of (7.13) with $(t - \theta)\partial_t u(t)$ we obtain

$$\int_\theta^t \|\partial_t u(\tau)\|^2 (\tau - \theta) \, d\tau \le M(\|u(\theta)\|), \quad \theta \le t \le \theta + 1, \quad (7.19)$$

where M does not explicitly depend on θ. And finally we differentiate (7.13) with respect to t, multiply both sides by $(t-\theta)\partial_t u$ and then integrate over t. We obtain

$$\|\partial_t u(t)\|^2 \le C_2(\|u(\theta)\|), \quad (7.20)$$

where $\theta + \delta \le t \le \theta + 1$, $\delta > 0$, $\forall \theta \ge 0$. For every fixed $t \ge \theta > 0$ we consider the equation (7.13) as an elliptic equation:

$$-\lambda \Delta^2 u(t) + \Delta u(t) = g_1(t), \quad (7.21)$$

where $g_1(t) = \partial_t u(t) + f(u(t)) + g$. From the inequality (7.20) and conditions imposed on f it follows that $g_1 \in L_2(\Omega)$. Therefore

$$\lambda^2 \|\Delta^2 u(t)\|^2 + \|\Delta u(t)\|^2 \le C\|g_1\|^2 \le C_1(\|u(0)\|).$$

Hence, in view of the interpolation inequality, we have

$$\lambda \|u(t)\|_3^2 \le C_2(\|u(0)\|) \quad \forall t \ge \delta > 0.$$

Integrating this inequality over t we obtain (7.16).

To prove now the inequality (7.14) we subtract the equality (7.15) from (7.13), take the scalar product of the result by $u - u_0$ and integrate over t. We obtain

$$\frac{1}{2}\|u(t) - u_0(t)\|^2 + \int_\theta^t \|\nabla u(\tau) - \nabla u_0(\tau)\|^2 \, d\tau$$

$$\le \frac{1}{2}\|u(\theta) - u_0(\theta)\|^2 - \int_\theta^t \langle f(u) - f(u_0), u - u_0 \rangle \, d\tau$$

$$+ \lambda \int_\theta^t \|u\|_3 \cdot \|u - u_0\|_1 \, d\tau,$$

that is

$$\|u(t) - u_0(t)\|^2$$

$$\leq C_1 \int_\theta^t \|u(\tau) - u_0(\tau)\|^2 \, d\tau + \|u(\theta) - u_0(\theta)\|^2 + \lambda M_1(|t - \theta| + 1).$$

Using the Gronwall inequality we have (7.14) ($\alpha = C_1$). From the estimate (7.14) and other properties of solutions of (7.13) and (7.14) follow the conditions of a theorem similar to Theorem 6.1. If B_0 is a bounded set of initial conditions for (7.13) in $L_2(\Omega)$ and $U_\lambda = \{u(t, \lambda) \mid u(0, \lambda) \in B_0\}$ is the corresponding set of solutions of (7.13) then for each $u(t, \lambda) \in U_\lambda$ there exists a c.l.t. $\tilde{u}(t, \lambda)$ ($\tilde{u}(t, \lambda)$ is a solution of (7.15)) such that

$$\sup_{t \geq \delta} \|u(t, \lambda) - \tilde{u}(t, \lambda)\| \leq C\lambda^q, \quad q > 0, \quad \delta > 0. \tag{7.22}$$

Here the constant C depends only on B_0, and q depends on spectra of the operators $\Delta v - f'(z_i)v$ at all equilibrium points

$$z_i \in \mathcal{M} \quad (i = 1, ..., N), \quad z_i|_{\partial\Omega} = 0.$$

Let us note finally that the main part of $\tilde{u}(t, \lambda)$ consists of pieces of curves which belong to the finitely-parametrized families of solutions of the equation (7.15) lying on $M^+(z_i)$ f($i = 1, ..., N$) (the first piece is the exception).

Remark 7.2.
The estimate (7.22) shows that $\tilde{u}(t, \lambda)$ is the chief member of the global (i.e. for all $t \geq \delta > 0$) asymptotics for $u(t, \lambda) = u_\lambda(t)$ in the metric $L_2(\Omega)$. We note that this metric does not take into account the existence of boundary layers. The reason is that boundary layer functions taking part in the asymptotic expansion of $u(t, \lambda)$ have $L_2(\Omega)$-norms which tend to zero if $\lambda \to 0$ (see Chapter VI and, for details, Vishik & Lusternik [1]).

V

The asymptotics of solutions of reaction-diffusion equations with small parameter

§8 Formulation of the problem

If the reaction-diffusion system depends on a large parameter μ,

$$\partial_t u_1 = a_1 \mu \Delta u_1 - \mu(f_1(u_1, u_2) + g_1(x)),$$
$$\partial_t u_2 = a_2 \Delta u_2 - f_2(u_1, u_2) - g_2(x),$$

it is possible, dividing by μ, to transform this system to another depending on a small parameter ε:

$$\varepsilon \partial_t u_1 = a_1 \Delta u_1 - f_1(u_1, u_2) - g_1(x), \tag{8.1}$$
$$\partial_t u_2 = a_2 \Delta u_2 - f_2(u_1, u_2) - g_2(x), \tag{8.2}$$

where $\varepsilon = 1/\mu$, $x \in \Omega \Subset \mathbf{R}^n$. Let $a_1 = 1$, $a_2 = 1$. We supplement this system with boundary conditions

$$u_1|_{\partial \Omega} = 0, \quad u_2|_{\partial \Omega} = 0 \tag{8.3}$$

or

$$\frac{\partial u_1}{\partial \nu}|_{\partial \Omega} = 0, \quad \frac{\partial u_2}{\partial \nu}|_{\partial \Omega} = 0 \tag{8.3'}$$

and the initial data

$$u_1|_{t=0} = u_{10}(x), \quad u_2|_{t=0} = u_{20}(x), \tag{8.4}$$

where $0 < \varepsilon \leq \varepsilon_0$, $g_i(x) \in L_2(\Omega)$ $(i = 1, 2)$. If some conditions on $f_1(\xi)$, $f_2(\xi)$ are satisfied, the problem (8.1) - (8.4) has a unique solution $u(t) = (u_1(t), u_2(t))$ $(u_i(t) \equiv u_i(t, x))$, which belongs to the space

$H_1 = H_1^0(\Omega) \times H_1^0(\Omega)$ for any $t \geq 0$. Therefore, this system generates a semigroup $\{S_t(\varepsilon)\}$, $S_t(\varepsilon) : H_1 \to H_1 \; \forall t \geq 0$, where $S_t(\varepsilon)u_0 = u(t)$ is a solution of the problem (8.1) - (8.4). For $\varepsilon = 0$ the system (8.1) - (8.4) becomes

$$0 = \Delta v_1 - f_1(v_1, v_2) - g_1(x), \tag{8.5}$$
$$\partial_t v_2 = \Delta v_2 - f_2(v_1, v_2) - g_2(x), \tag{8.6}$$
$$v_1|_{\partial\Omega} = 0, \quad v_2|_{\partial\Omega} = 0, \tag{8.7}$$
$$v_2|_{t=0} = v_{20}(x). \tag{8.8}$$

The equation (8.5) is a stationary one and owing to conditions on $f_1(v_1, v_2)$ (see §9) it has a unique solution $v_1(x) = Bv_2(x)$ for any function $v_2(x) \in H_1^0(\Omega)$. Substituting $v_1 = Bv_2$ in the second equation one can obtain an evolution equation for v_2. Let the semigroup $\{S_t \mid t \geq 0\}$, $S_t v_{20} = v_2(t)$, correspond to the second equation so obtained. It is supposed that the limit equation so obtained (for $\varepsilon = 0$) for v_2 has a finite number of equilibrium points and that all of them are hyperbolic.

Under the assumption that $(f_1(u), f_2(u))$ is a potential vector and $f_1(u)$, $f_2(u)$ satisfy some other conditions, the following statement is proved in §10.

Solutions $u(t, \varepsilon) = (u_1(t, \varepsilon), u_2(t, \varepsilon))$ of the problem (8.1) - (8.4) with the initial data from the bounded set $B_0 \subseteq H_1$ have a stabilized asymptotic, which is a solution $v(t) = (v_1(t), v_2(t))$ piecewise continuous with respect to t of the limit problem (8.5) - (8.8) and, moreover, $v(t)$ depends on ε $(v(t) = v(t, \varepsilon))$ and satisfies the estimates

$$\sup_{t \geq 0} \|u_2(t, \varepsilon) - v_2(t)\| \leq C_1 \varepsilon^q, \tag{8.9}$$
$$\sup_{t \geq \tau} \|u_1(t, \varepsilon) - v_1(t)\| \leq C_2 \varepsilon^q, \tag{8.10}$$

$q > 0$, $\tau > 0$. Here constants C_1 and C_2 depend on B_0, C_2 depends on τ also, q depends on spectral properties of the operators $S_1'(z_j)$, where z_j are equilibrium points, $S_1'(z_j) = S_t'(z_j)|_{t=1}$ is the Fréchet derivative of the operator S_t at the point z_j for $t = 1$. It is necessary to note that continuous pieces of $v_2(t)$, excepting the first one, lie on the finite-dimensional unstable manifolds of the limit semigroup $\{S_t\}$, passing through z_j.

So the trajectory $v(t) = (v_1(t), v_2(t))$ is the main term of asymptotics of $u(t, \varepsilon)$ for all $t \geq \tau$.

It is necessary to note that Dirichlet boundary conditions in (8.3) can be changed for Neumann conditions (8.3'). The proved theorems can be generalized to the case when u_1 and u_2 are vectors: $u_1 = (u_1^1, ..., u_1^m)$, $u_2 = (u_2^1, ..., u_2^m)$.

Finally, we should note that in the problem (8.1) - (8.4) the approximating trajectory $v(t)$ lies on M, where M is a manifold, determined by the stationary equation (8.5), and attracts the solutions $u(t, \varepsilon)$ of the prelimit equation. In the neighbourhood of $t = 0$ the component $u_1(t, \varepsilon)$ has the nature of a boundary layer (see, for instance, Vishik & Lusternik [1]) which is not investigated here (so $t \geq \tau$ in (8.10)).

We restrict ourselves to the case of Dirichlet boundary conditions:

$$u_1|_{\partial\Omega} = 0, \quad u_2|_{\partial\Omega} = 0.$$

Problems analogous to (8.1) - (8.4) have been investigated by many authors for systems of ordinary differential equations (see, e.g., Tichonov [1], [2]).

We write

$$u = (u_1, u_2), \quad f(u) = (f_1(u), f_2(u)), \quad f_i(u) \in C^{1+\alpha},$$
$$\alpha > 0 \ (i = 1, 2), \quad g(x) = (g_1(x), g_2(x)).$$

For $\varepsilon = 0$ (8.1) turns into the elliptic equation

$$\Delta u_1 - f_1(u_1, u_2) - g_1(x) = 0. \tag{8.11}$$

We shall assume that the function $f(u)$ satisfies the estimate

$$(f'(\xi)\eta, \eta) \geq \mu_1 \eta_1^2 + \mu_2 \eta_2^2, \quad \mu_1, \mu_2 \in \mathbf{R},$$
$$\mu_1 \geq -(\lambda_1 - \delta), \quad \forall \xi \in \mathbf{R}^2, \quad \forall \eta \in \mathbf{R}^2, \tag{8.12}$$

$\delta > 0$ is an arbitrary constant, and $-\lambda_1$ is the first eigenvalue of the Laplace operator with the boundary conditions (8.3) in the domain $\Omega \in \mathbf{R}^n$. The condition (8.12) implies, in particular, that

$$f'_{1\xi_1} \geq -(\lambda_1 - \delta), \tag{8.13}$$

$$f'_{2\xi_2} \geq \mu_2. \tag{8.14}$$

We shall assume also that the function $f(\xi)$ has the following properties:

$$(f(\xi), \xi) \geq -C + \delta_1^2 |\xi|^2, \quad f(0) = 0, \tag{8.15}$$

where δ_1^2 is sufficiently small,

$$|f'_{1\xi_2}(\xi)| \leq C(1+|\xi|^\eta), \quad \text{where } 1/\eta \geq (n-2)/2. \tag{8.16}$$

We assumed (8.15) only for simplicity. Actually the result is still valid when

$$(f(\xi),\xi) \geq -C - \delta_2^2|\xi|^2, \quad f(0) = 0.$$

Moreover, let $f(\xi)$ be a potential vector, that is there exists such a function $F(\xi_1, \xi_2)$, that

$$f(\xi) = \text{grad } F(\xi). \tag{8.17}$$

It is supposed that

$$-C \leq F(\xi) \leq C(1+|\xi|^{p+1}), \quad \text{where } 1/p \geq (n-2)/n. \tag{8.18}$$

We should note that the conditions (8.16), (8.17) imply

$$|f'_{2\xi_1}(\xi)| \leq C(1+|\xi|^\eta) \tag{8.19}$$

because

$$\frac{\partial^2 F}{\partial \xi_1 \partial \xi_2} = \frac{\partial^2 F}{\partial \xi_2 \partial \xi_1}.$$

We shall assume also that

$$|f(\xi)| \leq C(1+|\xi|^p). \tag{8.20}$$

It will be proved below that the equation (8.11) has a unique solution $u_1(x) \in H_1^0(\Omega)$ for any function $u_2(x) \in H_1^0(\Omega)$, $u_1 = B(u_2)$, when $f(\xi)$ satisfies the conditions (8.12) - (8.20).

Moreover, let the function $g_2(x) \in L_2(\Omega)$ be such that
(1) a set $\mathcal{M}$ of solutions of the equation

$$\Delta u_2 - f_2(B(u_2), u_2) - g_2(x) = 0$$

is finite, $\mathcal{M} = \{z_1, ..., z_N\}$,
(2) all $z_j \in \mathcal{M}$ are hyperbolic (see §2).

The existence of a unique solution of the problem (8.1) - (8.4) is proved using Galerkin's method (see, for instance, Babin & Vishik [1], Lions [1]). The rigorous conclusion of the computations, done below, can be proved simply with Galerkin's method also. So we restrict ourselves only to the formal reasonings.

§9   A priori estimates

We shall write $H = (L_2(\Omega))^2$, $H_1 = (H_1^0(\Omega))^2$, $\{S_t(\varepsilon)\}$ is a semigroup, corresponding to the problem (8.1) - (8.4). Operators $S_t(\varepsilon)$ map the initial data $(u_{10}(x), u_{20}(x)) = u_0(x)$ into the solution $(u_1(t,x), u_2(t,x)) = u(t,x)$ of the problem (8.1) - (8.4) at the moment t: $S_t(\varepsilon)u_0 = u(t)$ $(u(t) \equiv u(t,x),\ u_0 \equiv u_0(x))$.

First we establish some estimates for the component $u_2(t)$.

Lemma 9.1.

Let $u(t) = (u_1(t), u_2(t))$ be a solution of the problem (8.1) - (8.4), $0 < \varepsilon \le 1$. Then there exist a constant $C_1 > 0$ and a function $C_2(\cdot) > 0$, independent of ε, such that

(1) $\|u_2(t)\|^2 \le C_1(1 + \|u(0)\|^2)$ $\forall t \ge 0$,

(2) $\|u_2(t)\|_1^2 \le C_2(\|u(0)\|_1)$ $\forall t \ge 0$.

Proof   (1) We multiply the system (8.1) - (8.2) by u in H and use the condition (8.15). Then we get

$$\frac{1}{2}(\varepsilon\partial_t\|u_1\|^2 + \partial_t\|u_2\|^2) + \|\nabla u\|^2 = -(f(u), u) - (g, u)$$

$$\le C - \delta_1^2\|u\|^2 + M\|g\|^2 + \frac{1}{2}\delta_1^2\|u\|^2$$

$$\le C' - \frac{1}{2}\delta_1^2\|u\|^2.$$

$$(9.1)$$

Multiplying both sides of the inequality (9.1) by $e^{\delta_1^2 t}$, where δ_1^2 is sufficiently small, and integrating it from 0 to t we deduce

$$\varepsilon e^{\delta_1^2 t}\|u_1(t)\|^2 + e^{\delta_1^2 t}\|u_2(t)\|^2 + \delta_1^2(1-\varepsilon)\int_0^t e^{\delta_1^2 \tau}\|u_1(\tau)\|^2\,d\tau$$

$$\le C''(e^{\delta_1^2 t} - 1) + \varepsilon\|u_1(0)\|^2 + \|u_2(0)\|^2.$$

Hence

$$\|u_2(t)\|^2 \le C(1 - e^{-\delta_1^2 t}) + \|u(0)\|^2 e^{-\delta_1^2 t}$$

and statement (1) is proved.

(2) Integrating inequality (9.1) from 0 to t we get

$$\varepsilon\|u_1(t)\|^2 + \delta_1^2\int_0^t \|u_1(\tau)\|^2\,d\tau \le C't + \varepsilon\|u_1(0)\|^2 + \|u_2(0)\|^2.$$

As $||u_1(t)||^2 \geq 0$, $||u_1(t)||^2 \leq y(t)$, where $y(t)$ satisfies the equation

$$\varepsilon y(t) + \delta_1^2 \int_0^t y(\tau)\,d\tau = C''(1+t), \quad C'' = C' + ||u(0)||^2,$$

or

$$\varepsilon y'(t) + \delta_1^2 y(t) = C'', \quad y(0) = \frac{1}{\varepsilon}C''.$$

Therefore,

$$||u_1(t)||^2 \leq y(t) = y(0)e^{-\delta_1^2 t/\varepsilon} + C_1(1 - e^{-\delta_1^2 t/\varepsilon})$$
$$\leq \frac{C''}{\varepsilon}e^{-\delta_1^2 t/\varepsilon} + C_1(1 - e^{-\delta_1^2 t/\varepsilon}).$$

Since $xe^{-x} \leq C_0$,

$$\frac{1}{\varepsilon}e^{-\delta_1^2 t/\varepsilon} \leq \frac{C_0}{\delta_1^2} \cdot \frac{1}{t}.$$

Hence

$$||u_1(t)||^2 \leq M'(1 + ||u(0)||^2) \quad \forall t \geq 1. \tag{9.2}$$

Multiplying the system (8.1) - (8.2) by $-t\Delta u$ in H, we conclude

$$\frac{1}{2}(\varepsilon\partial_t(t||\nabla u_1||^2) + \partial_t(t||\nabla u_2||^2)) + t||\Delta u||^2$$
$$= \frac{1}{2}\left(\varepsilon||\nabla u_1||^2 + ||\nabla u_2||^2\right) - t(f'(u)\nabla u, \nabla u) + t(g, \Delta u)$$
$$\leq \frac{1}{2}(\varepsilon||\nabla u_1||^2 + ||\nabla u_2||^2)$$
$$+ t(\lambda_1 - \delta)||\nabla u_1||^2 - \mu_2 t||\nabla u_2||^2$$
$$+ Mt||g||^2 + \delta_2 t||\Delta u||^2. \tag{9.3}$$

Let now $\delta_2 > 0$ be such that

$$(1 - \delta_2)||\Delta u||^2 \geq \left(\lambda_1 - \frac{\delta}{2}\right)||\nabla u||^2. \tag{9.4}$$

One can see also that integrating the inequality (9.1) from 0 to t we get

$$\int_0^t ||\nabla u(\tau)||^2\,d\tau \leq C(1+t), \quad C = C(||u(0)||). \tag{9.5}$$

The estimates (9.3) - (9.5) imply that

$$\varepsilon t||\nabla u_1(t)||^2 + t||\nabla u_2(t)||^2 + \delta \int_0^t \tau||\nabla u_1(\tau)||^2 \, d\tau$$

$$\leq \int_0^t \left(\varepsilon||\nabla u_1(\tau)||^2 + ||\nabla u_2(\tau)||^2\right) \, d\tau - \mu_2 \int_0^t \tau||\nabla u_2(\tau)||^2 \, d\tau + Ct^2$$

$$\leq -\mu_2 \int_0^t \tau||\nabla u_2(\tau)||^2 \, d\tau + C'(1 + T^2)$$

for $0 \leq t \leq T$. Hence we get, using the Gronwall inequality, for t, $0 \leq t \leq T$,

$$t||\nabla u_2(t)||^2 \leq C, \quad C = C(T, ||u(0)||^2). \tag{9.6}$$

Now, multiplying the system (8.1) - (8.2) by $-\Delta u$ in H, we obtain

$$\frac{1}{2}(\varepsilon\partial_t||\nabla u_1||^2 + \partial_t||\nabla u_2||^2) + ||\Delta u||^2 = -(f'(u)\nabla u, \nabla u) + (g, \Delta u)$$

$$\leq (\lambda_1 - \delta)||\nabla u_1||^2 - \mu_2||\nabla u_2||^2 + M||g||^2 + \delta_2||\Delta u||^2,$$

where δ_2 satisfies the condition (9.4). Therefore,

$$\frac{1}{2}(\varepsilon\partial_t||\nabla u_1||^2 + \partial_t||\nabla u_2||^2) \leq -\mu_2||\nabla u_2||^2 + C.$$

Integrating over $[0, t]$ we get

$$||\nabla u_2(t)||^2 \leq C_1 \text{ for } 0 \leq t \leq T, \quad C_1 = C_1(T, ||\nabla u(0)||). \tag{9.7}$$

Then we should note that item (1) of this lemma and the formulas (9.2) and (9.6) imply

$$||\nabla u_2(t)||^2 \leq C(||u(t-1)||) \leq C_1(||u(0)||) \text{ for } t \geq 2 \tag{9.8}$$

and the formula (9.7) implies

$$||\nabla u_2(t)||^2 \leq C(||\nabla u(0)||) \text{ for } 0 \leq t \leq 2. \tag{9.9}$$

Item (2) follows from the estimates (9.8) and (9.9).

Now we establish some properties of the semigroup $\{S_t(\varepsilon)\}$.

Lemma 9.2.

The semigroups $\{S_t(\varepsilon)\}$ *are*

(1) (H_1, H)*-bounded uniformly with respect to* ε, $0 < \varepsilon \leq \varepsilon_0$ *and* $t \geq 0$,
that is for any bounded set B_1 *in* H_1 *there exists a set* B_0, *bounded
in* H, *independent of* ε *and* t, $0 < \varepsilon \leq \varepsilon_0$, $0 \leq t < +\infty$, *such that*
$S_t(\varepsilon)B_1 \subseteq B_0$,

(2) (H_1, H_1)*-bounded uniformly with respect to* ε, $0 < \varepsilon \leq \varepsilon_0$, *and* $t \geq 0$.

Proof   (1) Multiplying the equation (8.1) by u_1 in $L_2(\Omega)$ we get

$$\frac{1}{2}\varepsilon\partial_t\|u_1\|^2 = -\|\nabla u_1\|^2 - (f_1(u_1, u_2), u_1) - (g_1, u_1).$$

One can see that

$$f_1(u_1, u_2) = f_1(0, u_2) + \int_0^t f'_{1u_1}(\theta u_1, u_2)\, \mathrm{d}\theta \cdot u_1,$$

and we get from (8.3) that

$$\frac{1}{2}\varepsilon\partial_t\|u_1\|^2 \leq -\lambda_1\|u_1\|^2 - (f_1(0, u_2), u_1) + (\lambda_1 - \delta)\|u_1\|^2$$
$$+ M\|g_1\|^2 + \frac{\delta}{4}\|u_1\|^2.$$

Then using condition (8.20) we obtain

$$\begin{aligned}
\frac{1}{2}\varepsilon\partial_t\|u_1\|^2 &\leq -\frac{\delta}{2}\|u_1\|^2 + C_1\|f_1(0, u_2)\|^2 + M\|g_1\|^2 \\
&\leq -\frac{\delta}{2}\|u_1\|^2 + C(1 + \|u_2\|_{0,2p}^{2p}) \qquad (9.10) \\
&\leq -\frac{\delta}{2}\|u_1\|^2 + C(1 + \|u_2\|_1^{2p}).
\end{aligned}$$

Lemma 9.1 and inequality (9.10) imply that

$$\varepsilon\partial_t\|u_1\|^2 + \delta\|u_1\|^2 \leq C_2, \quad C_2 = C_2(\|u(0)\|_1).$$

Therefore

$$\|u_1(t)\|^2 \leq C_3(1 - e^{-\delta t/\varepsilon}) + \|u(0)\|^2 e^{-\delta t/\varepsilon}, \quad C_3 = C_3(\|u(0)\|_1)$$

and taking into account Lemma 9.1 we get statement (1).

(2) We need now the following functional:

$$P(u) = \int_\Omega \left(\frac{1}{2}|\nabla u|^2 + F(u) + g \cdot u\right) dx, \quad u \in H_1, \tag{9.11}$$

where $F(u)$ is determined by (8.17). (This functional is a global Lyapunov function of the problem (8.1), (8.2).)

We should note that if $u(t)$ is a trajectory of the system (8.1) - (8.2) then

$$\partial_t P(u(t)) = -\varepsilon \|\partial_t u_1\|^2 - \|\partial_t u_2\|^2. \tag{9.12}$$

It follows from the equality (9.12) that

$$P(u(t)) \leq P(u(0)), \quad \forall t \geq 0. \tag{9.13}$$

The condition (8.18) implies

$$\begin{aligned}
\frac{1}{2}\|\nabla u\|^2 - C - \|g\|^2 - \|u\|^2 &\leq P(u) \\
&\leq \frac{1}{2}\|\nabla u\|^2 + C(1 + \|u\|_{0,1+p}^{1+p}) + \|g\|^2 + \|u\|^2.
\end{aligned} \tag{9.14}$$

Now using inequalities (9.13) and (9.14) we obtain

$$\|\nabla u(t)\|^2 \leq 2\|u(t)\|^2 + 2\|u(0)\|_1^2 + M(1 + \|u(0)\|_1^{1+p}).$$

Hence and from item (1) of this Lemma statement (2) follows.

Now we estimate the so-called integral of dissipation, that is the integral with respect to t of $\varepsilon\|\partial_t u_1\|^2 + \|\partial_t u_2\|^2$.

Lemma 9.3.

Let $u(t) = (u_1(t), u_2(t))$ be a solution of the problem (8.1) - (8.4). Then uniformly with respect to ε, $0 < \varepsilon \leq \varepsilon_0$, the following estimates hold:

$$\int_0^\infty \left(\varepsilon\|\partial_t u_1(\tau)\|^2 + \|\partial_t u_2(\tau)\|^2\right) d\tau \leq C_1, \quad C_1 = C_1(\|u(0)\|_1), \tag{9.15}$$

$$t(\varepsilon\|\partial_t u_1(t)\|^2 + \|\partial_t u_2(t)\|^2) \leq C_2(1+t), \quad C_2 = C_2(\|u(0)\|_1). \tag{9.16}$$

Proof   First we prove the estimate (9.15). (9.12) and (9.14) imply

$$\int_0^t \left(\varepsilon ||\partial_t u_1(\tau)||^2 + ||\partial_t u_2(\tau)||^2 \right) \mathrm{d}\tau = P(u(0)) - P(u(t))$$
$$\leq ||u(t)||^2 + M(1 + ||u(0)||_1^{1+p}) + ||u(0)||_1^2.$$

Hence we obtain, using Lemma 9.2,

$$\int_0^t \left(\varepsilon ||\partial_t u_1(\tau)||^2 + ||\partial_t u_2(\tau)||^2 \right) \mathrm{d}\tau \leq C_1, \quad C_1 = C_1(||u(0)||_1).$$

Since C_1 does not depend on t we get (9.15).

Then we prove the estimate (9.16). If we differentiate the system (8.1) - (8.2) with respect to t and write the derivative $\partial_t u_i$ as u_i' $(i = 1, 2)$ we can deduce

$$\varepsilon \partial_t u_1' = \Delta u_1' - f_{1u_1}'(u)u_1' - f_{1u_2}'(u)u_2',$$
$$\partial_t u_2' = \Delta u_2' - f_{2u_1}'(u)u_1' - f_{2u_2}'(u)u_2'.$$

Multiplying this system by tu' in H we get

$$\frac{\varepsilon}{2}\partial_t(t||u_1'||^2) + \frac{1}{2}\partial_t(t||u_2'||^2) + t||\nabla u'||^2$$
$$= \frac{\varepsilon}{2}||u_1'||^2 + \frac{1}{2}||u_2'||^2 - t(f'(u)u', u').$$

Then we integrate the equality obtained from 0 to t and use the condition (8.12):

$$\varepsilon t ||\partial_t u_1||^2 + t||\partial_t u_2||^2 + \int_0^t \tau ||\nabla \partial_t u(\tau)||^2 \mathrm{d}\tau$$
$$\leq (\lambda_1 - \delta) \int_0^t \tau ||\partial_t u_1(\tau)||^2 \mathrm{d}\tau - \mu_2 \int_0^t \tau ||\partial_t u_2(\tau)||^2 \mathrm{d}\tau$$
$$+ \int_0^t \frac{1}{2}(\varepsilon ||\partial_t u_1(\tau)||^2 + ||\partial_t u_2(\tau)||^2) \mathrm{d}\tau.$$

We should note that $||\nabla v||^2 \geq \lambda_1 ||v||^2$. Moreover taking into account the estimate (9.15) we get

$$t(\varepsilon ||\partial_t u_1(t)||^2 + ||\partial_t u_2(t)||^2) \leq |\mu_2| t C_1 + C_1 \leq C_2(1 + t),$$

$C_2 = C_2(||u(0)||_1)$.

Corollary 9.1.
From (9.16) it follows that

$$\varepsilon||\partial_t u_1(t)||^2 + ||\partial_t u_2(t)||^2 \leq C_3 = C_3(||u(0)||_1, \tau) \quad \text{for } t \geq \tau > 0. \quad (9.17)$$

§10   The stabilized asymptotics of $u(t)$

We need the following lemma.

Lemma 10.1
If $|\phi(\xi)| \leq C(1 + |\xi|^\eta)$, $1/\eta \geq (n-2)/2$, then for any $u, w \in H_1$, $v \in L_2$, $\alpha > 0$ there exists $C > 0$ such that the following estimate holds:

$$||\phi(u(x))v(x)w(x)||_{0,1} \leq C(1 + ||u||_1^{2\eta})||v||^2 + \alpha||w||_1^2.$$

Proof   One can see that the Hölder inequality implies

$$||\phi(u)vw||_{0,1} \leq ||\phi(u)||_{0,p_1}||v||_{0,p_2}||w||_{0,p_3}, \quad \text{where } \frac{1}{p_1} + \frac{1}{p_2} + \frac{1}{p_3} = 1. \quad (10.1)$$

We restrict ourselves to proving the lemma in the case $n > 2$.
Let $p_1 = n$, $p_2 = 2$, $p_3 = 2n/(n-2)$. Then

$$||w||_{0,p_3}^2 \leq C_1||w||_1^2.$$

Moreover,

$$||\phi(u)||_{0,p_1} \leq C_2(1 + ||u||_{0,\eta \cdot n}^\eta) \leq C_3(1 + ||u||_1^\eta).$$

From these estimates and from (10.1) it follows that

$$||\phi(u)vw||_{0,1} \leq C_4||\phi(u)||_{0,p_1}^2||v||_{0,p_2}^2 + \alpha_1||w||_{0,p_3}^2$$
$$\leq C_5(1 + ||u||_1^{2\eta})||v||^2 + \alpha_1||w||_1^2.$$

As $\alpha_1 > 0$ is arbitrary the lemma is proved.
Now we formulate Theorem 6.1' in the notation we shall use below.

Theorem 10.1.

Let $\{S_t\}$ be a semigroup acting in E, where E is a Banach space, and B_0 be a bounded set in E, $S_t(B_0) \subseteq B_0 \; \forall t \geq 0$. Let U_ε be a set of functions $u(\cdot, \varepsilon) = u(\cdot) \in U_\varepsilon$ dependent on a parameter ε, such that the set of values of these functions

$$B_\varepsilon = \{u = u(t) \mid u(t) \in U_\varepsilon\}$$

lies in E, $B_\varepsilon \subseteq E$. We suppose that the following conditions hold.

(1) A set $\mathcal{M}$ of equilibrium points of $\{S_t\}$ is finite, $\mathcal{M} = \{z_1, ..., z_N\}$, $\mathcal{M} \subseteq B_0$.

(2) $\forall \delta > 0 \; \exists T^0 < +\infty$ such that T^0 is a time of arrival from B_0 to $\mathcal{O}_\delta(\mathcal{M})$.

(3) $\forall u_0 \in B_0$ there exist an enumeration $\{z_j\}$ of points $z_j \in \mathcal{M}$ and $\delta > 0$ such that $S_t u_0$ passes $\mathcal{O}_\delta(z_j)$ in reverse order to this enumeration. Moreover, it is supposed that $\{S_t\}$ has a global Lyapunov function.

(4) $\forall j = 1, ..., N \; z_j \in \mathcal{M}$ is a hyperbolic equilibrium point.

(5) $B_\varepsilon \subseteq B_0 \; \forall \varepsilon \in (0, \varepsilon_1]$.

(6) There exist positive constants α, C, q_0 such that $\forall u(t) \in U_\varepsilon$, $\forall v_0 \in B_0$, $\forall t, \tau, \; t \geq \tau > 0$,

$$\|u(t) - S_{t-\tau} v_0\|_E \leq C e^{\alpha(t-\tau)}(\varepsilon^{q_0} + \|u(\tau) - v_0\|_E).$$

(7) $u(t) \in U_\varepsilon$ is continuous in t for any $t \geq 0$.

Then there exist $\varepsilon_0 > 0$, $q > 0$, $C > 0$ such that $\forall \varepsilon \in (0, \varepsilon_0]$, $\forall u(t) \in U_\varepsilon$ there exists a combined limit trajectory $\tilde{u}(t, \varepsilon)$ (see Definition 6.2) satisfying the inequality

$$\sup_{t \geq 0} \|u(t) - \tilde{u}(t, \varepsilon)\|_E \leq C \varepsilon^q. \tag{10.1'}$$

To prove that the conditions of Theorem 10.1 hold for the solutions $u(t)$ of the system (8.1) - (8.4) and the solutions $v(t)$ of the system (8.5) - (8.7) we need some auxiliary propositions. We should recall that $v_1 = B v_2$ is a solution of the stationary equation (8.5), moreover $B : H_1^0(\Omega) \to H_1^0(\Omega)$ (see Lemma 10.3 below). The existence of the operator B follows from the properties of the function $f_1(v_1, v_2)$.

Lemma 10.2.

Let $u(t) = (u_1(t), u_2(t)) \in H_1$ be a solution of the problem (8.1) - (8.4). Then

(1) $\|u_1(t) - B(u_2(t))\|_1^2 \leq \varepsilon C_1,$ $C_1 = C_1(\|u(0)\|_1, \tau),$ $\forall t \geq \tau > 0,$

(2) $\displaystyle\int_0^t \|u_1(\tau) - B(u_2(\tau))\|_1^2 \, d\tau \leq \varepsilon C_2,$ $C_2 = C_2(\|u(0)\|_1),$ $\forall t \geq 0.$

Proof   Subtracting equation (8.11), where u_1 is replaced by $B(u_2)$, from equation (8.1) we get

$$\varepsilon \partial_t u_1 = \Delta(u_1 - B(u_2)) - (f_1(u_1, u_2) - f_1(B(u_2), u_2)). \qquad (10.2)$$

Multiplying equation (10.2) by $u_1 - B(u_2)$ in $L_2(\Omega)$ and using the condition (8.13) we obtain

$$0 = -\|\nabla(u_1 - B(u_2))\|^2$$
$$- (f_1(u_1, u_2) - f_1(B(u_2), u_2), u_1 - B(u_2)) - (\varepsilon \partial_t u_1, u_1 - B(u_2))$$
$$\leq -\delta_0^2 \|\nabla(u_1 - B(u_2))\|^2 - \left(\lambda_1 - \frac{\delta}{4}\right) \|u_1 - B(u_2)\|^2$$
$$- \left(\int_0^1 f'_{1u_1}(B(u_2) + \theta(u_1 - B(u_2)), u_2) \, d\theta(u_1 - B(u_2)), u_1 - B(u_2)\right)$$
$$+ M\varepsilon^2 \|\partial_t u_1\|^2 + \frac{\delta}{4} \|u_1 - B(u_2)\|^2$$
$$\leq -\delta_0^2 \|\nabla(u_1 - B(u_2))\|^2 - \frac{\delta}{2} \|u_1 - B(u_2)\|^2 + M\varepsilon^2 \|\partial_t u_1\|^2.$$

We have used in this reasoning that

$$-\left(\int_0^1 f'_{1u_1}(B(u_2) + \theta(u_1 - B(u_2)), u_2) \, d\theta(u_1 - B(u_2)), u_1 - B(u_2)\right)$$
$$\leq (\lambda_1 - \delta)\|u_1 - B(u_2)\|^2.$$

Hence
$$\|u_1 - B(u_2)\|_1^2 \leq M'\varepsilon^2 \|\partial_t u_1\|^2. \qquad (10.3)$$

Now we deduce from the estimates (9.17) and (10.3) that

$$\|u_1(t) - B(u_2(t))\|_1^2 \leq \varepsilon M' \varepsilon \|\partial_t u_1\|^2 \leq \varepsilon C_1, \quad C_1 = C_1(\|u(0)\|_1, \tau),$$

for $t \geq \tau > 0$, and from the estimates (9.15) and (10.3) that

$$\int_0^t \|u_1(\tau) - B(u_2(\tau))\|_1^2 \, d\tau \leq \varepsilon M' \int_0^t \varepsilon \|\partial_t u_1\|^2 \, d\tau \leq \varepsilon C_2,$$

where $C_2 = C_2(\|u(0)\|_1)$, $\forall t \geq 0$.

Lemma 10.3.
    Let $v(x) = (v_1(x), v_2(x))$, $w(x) = (w_1(x), w_2(x))$ be solutions of the equation (8.11) ($v_1 = B(v_2)$, $w_1 = B(w_2)$). Then

$$\|B(v_2) - B(w_2)\|_1 \leq M\|v_2 - w_2\|, \text{ where } M = M(\|v_2\|_1, \|w_2\|_1).$$

Proof   We have

$$\Delta(v_1 - w_1) - (f_1(v_1, v_2) - f_1(w_1, w_2)) = 0. \tag{10.4}$$

Multiplying equation (10.4) by $v_1 - w_1$ in $L_2(\Omega)$ we get

$$0 = -\|\nabla(v_1 - w_1)\|^2$$
$$- \left(\int_0^1 f'_{1u_1}(w_1 + \theta(v_1 - w_1), v_2) \, d\theta(v_1 - w_1), v_1 - w_1 \right)$$
$$- \left(\int_0^1 f'_{1u_2}(w_1, w_2 + \theta(v_2 - w_2)) \, d\theta(v_2 - w_2), v_1 - w_1 \right).$$

Now using the conditions (8.13), (8.16) and Lemma 10.1 we obtain

$$0 \leq -\delta_0^2 \|\nabla(v_1 - w_1)\|^2 - \left(\lambda_1 - \frac{\delta}{2} \right) \|v_1 - w_1\|^2 + (\lambda_1 - \delta)\|v_1 - w_1\|^2$$
$$+ C \left(1 + \|v_2\|_1^{2\eta} + \|w\|_1^{2\eta} \right) \|v_2 - w_2\|^2 + \frac{1}{2}\delta_0^2 \|v_1 - w_1\|_1^2, \tag{10.4'}$$

$\delta > \delta_0^2 > 0$. Multiplying the equation (8.11), where $(u_1, u_2) = (w_1, w_2)$, by w_1 and using the identity $f_1(w_1, w_2) = (f(w_1, w_2) - f(0, w_2)) + f(0, w_2)$ plus conditions (8.13), (8.20) we obtain

$$\|w_1\|_1^2 \leq M_1(\|w_2\|_1).$$

Then $\|w\|_1^{2\eta} \leq M_2(\|w_2\|_1)$. Hence, and from (10.4') we deduce that

$$\|v_1 - W_1\|_1^2 \leq (M(\|v_2\|_1, \|w_2\|_1))^2 \|v_2 - w_2\|^2.$$

We should note that $v_1 = B(v_2)$, $w_1 = B(w_2)$ and we get a complete proof of the lemma.

Now we use Theorem 10.1 for constructing the asymptotics of $u_2(t)$. Let $E = L_2(\Omega)$,

$$U_\varepsilon = \{u_2(t) \mid u(t) = (u_1(t), u_2(t)) \text{ solves } (8.1) - (8.4), \|u(0)\|_1 \leq M\}.$$

Let a semigroup $\{S_t\}$ correspond to the following problem:

$$\partial_t v_2 = \Delta v_2 - f_2(B(v_2), v_2) - g_2(x), \tag{10.5}$$

$$v_2|_{\partial\Omega} = 0, \quad v_2|_{t=0} = v_{20}(x), \quad x \in \Omega. \tag{10.6}$$

The semigroup $\{S_t\}$ acts in $L_2(\Omega)$, $S_t : L_2(\Omega) \to L_2(\Omega)$, $\forall t \geq 0$. This fact is established analogously to the proof of Lemma 9.1. Let

$$B_0 = \cup_{t \geq 0} S_t(\{u \in L_2(\Omega) \mid \|u\|_1 \leq M_1\}).$$

As $\{S_t\}$ is (H_1, H_1)-bounded uniformly with respect to $t \geq 0$, B_0 is bounded in H_1 and $S_t(B_0) \subseteq B_0$ $\forall t \geq 0$.

Let us now check all the conditions of Theorem 10.1.

We should note that it was supposed in §8 that conditions (1) and (4) of Theorem 10.1 hold.

Condition (2) follows from the boundedness of the integral of dissipation (formula (9.15)) and the boundedness of $u_2(t)$ in $H_1(\Omega)$ (see Babin & Vishik [1], page 257, or proofs of Theorem 5.2 and (5.18)).

The Lyapunov function looks like

$$P_1(u_2) =$$

$$\int_\Omega \left\{ \frac{1}{2}|\nabla B(u_2)|^2 + \frac{1}{2}|\nabla u_2|^2 + F(B(u_2), u_2) + g_1 \cdot B(u_2) + g_2 \cdot u_2 \right\} \, \mathrm{d}x.$$

Indeed it is easy to see, using the system (8.1) - (8.2) when $\varepsilon = 0$, that

$$\partial_t P_1(u_2(t)) = -\int_\Omega \partial_t u_2 \cdot \partial_t u_2 \, \mathrm{d}x < 0. \tag{10.7}$$

Here $u_2(t)$ is any solution of the equation (10.5) different from the stationary one. To deduce (10.7) one should use that $u_1(t) = B(u_2(t), u_2(t))$ is a solution of the system (8.1) - (8.2) for $\varepsilon = 0$.

Condition (3) follows from the existence of the Lyapunov function, the hyperbolicity of the equilibrium points and the precompactness of the set $\{S_t v_{20} \mid t \geq 1\}$ in $L_2(\Omega)$, which follows from Lemma 9.1, valid also for $\varepsilon = 0$.

Condition (5) holds owing to Lemma 9.1.

Condition (7) follows from the finiteness of the integral of dissipation (formula (9.15)) and the boundedness of $u_2(t) \in U_\varepsilon$ in $L_2(\Omega)$ (Lemma 9.1).

Now we move on to checking condition (6) of Theorem 10.1. Subtracting equation (10.5) from equation (8.2) we obtain

$$\partial_t(u_2 - v_2) = \Delta(u_2 - v_2) - (f_2(u_1, u_2) - f_2(B(v_2), v_2)). \qquad (10.8)$$

Multiplying (10.8) by $u_2 - v_2$ in $L_2(\Omega)$ and using the conditions (8.14), (8.19) and Lemma 10.1 we deduce

$$\frac{1}{2}\partial_t \|u_2 - v_2\|^2 + \|\nabla(u_2 - v_2)\|^2$$

$$= -\left(\int_0^1 f'_{2u_1}(B(v_2) + \theta(u_1 - B(v_2)), u_2) \, d\theta(u_1 - B(v_2)), u_2 - v_2 \right)$$

$$- \left(\int_0^1 f'_{2u_2}(B(v_2), v_2 + \theta(u_2 - v_2)) \, d\theta(u_2 - v_2), u_2 - v_2 \right)$$

$$\leq C \left(1 + \|u\|_1^{2\eta} + \|B(v_2)\|_1^{2\eta} \right) \|u_1 - B(v_2)\|^2$$

$$+ \frac{1}{2}\|u_2 - v_2\|_1^2 + \mu_2 \|u_2 - v_2\|^2.$$

$$(10.9)$$

We should note that since $\|u(t)\|_1$, $\|B(v_2)\|_1$ are bounded uniformly with respect to $t \geq 0$ (Lemma 9.2)

$$C \left(1 + \|u\|_1^{2\eta} + \|B(v_2)\|_1^{2\eta} \right) \leq M.$$

Therefore from (10.9) we get, integrating it from 0 to t,

$$\|u_2(t) - v_2(t)\|^2 \leq \|u_2(0) - v_2(0)\|^2 + M_1 \int_0^t \|u_1(\tau) - B(u_2(\tau))\|^2 \, d\tau$$

$$+ M_1 \int_0^t \|B(u_2(\tau)) - B(v_2(\tau))\|^2 \, d\tau$$

$$+ 2\mu_2 \int_0^t \|u_2(\tau) - v_2(\tau)\|^2 \, d\tau.$$

Using Lemmas 10.2 and 10.3 we obtain

$$\|u_2(t) - v_2(t)\|^2 \leq \|u_2(0) - v_2(0)\|^2 + M_2 \int_0^t \|u_2(\tau) - v_2(\tau)\|^2 \, d\tau + M_2\varepsilon.$$

Hence owing to the Gronwall inequality condition (6) has been proved.

Therefore we have checked all the conditions of Theorem 10.1 and the following theorem is valid.

Theorem 10.2.

For any set B_0 bounded in H_1 and any $\tau > 0$ there exist positive constants C_1, C_2, q, ε_0 such that for any solution $u(t) = (u_1(t), u_2(t))$ of the problem (8.1) - (8.4) with $0 < \varepsilon \leq \varepsilon_0$ and initial data from B_0 there exists a c.l.t. $\tilde{u}_2(t, \varepsilon)$, corresponding to the problem (8.5) - (8.6), which satisfies the following estimates:

$$\sup_{t \geq 0} \|u_2(t) - \tilde{u}_2(t, \varepsilon)\| \leq C_1 \varepsilon^q, \tag{10.10}$$

$$\sup_{t \geq \tau} \|u_1(t) - B(\tilde{u}_2(t, \varepsilon))\|_1 \leq C_2 \varepsilon^{q'}, \quad q' = \min(q, 1/2). \tag{10.11}$$

Proof   The estimate (10.10) holds owing to Theorem 10.1. We will show that the estimate (10.11) holds. We have

$$\|u_1(t) - B(\tilde{u}_2(t, \varepsilon))\|_1^2$$
$$\leq \|u_1(t) - B(u_2(t))\|_1^2 + \|B(u_2(t)) - B(\tilde{u}_2(t, \varepsilon))\|_1^2.$$

Using Lemmas 10.2 and 10.3 and the estimate (10.10) we obtain

$$\|u_1(t) - B(\tilde{u}_2(t, \varepsilon))\|_1^2 \leq C_1' \varepsilon + M^2 \|u_2(t) - \tilde{u}_2(t, \varepsilon)\|^2$$
$$\leq C_1' \varepsilon + C_1'' \varepsilon^{2q}$$

for $t \geq \tau > 0$. Hence the estimate (10.11) follows.

VI

The asymptotics of elements lying on the attractor of solutions of the perturbed evolutionary equations

§11   The main result

We shall investigate singular perturbed systems:

$$\frac{\partial u}{\partial t} = -\varepsilon^2 \Delta^2 u + \Delta u - f(u) - g(x), \quad x \in \Omega, \quad \Omega \in \mathbf{R}^n,$$
$$u = (u^1, ..., u^m), \quad f(u) = (f^1(u), ..., f^m(u)), \quad g(x) = (g^1(x), ..., g^m(x)).$$
$$(11.1)$$

We suppose that $f(u)$ satisfies some conditions. For $\varepsilon = 0$ system (11.1) coincides with a reaction-diffusion system. For simplicity we give the exposition for the scalar case $m = 1$. Generalization for arbitrary m is evident.

Asymptotic expansions of solutions of singularly perturbed evolution equations on a finite interval $[0, T]$ were investigated in many papers (see, for instance, Vishik & Lusternik [1], [2]). The solution $u^\varepsilon(x, t) \equiv u$ of the boundary value problem (11.1),

$$u|_{\partial\Omega} = 0, \quad \frac{\partial u}{\partial \nu}\big|_{\partial\Omega} = 0, \tag{11.2}$$

$$u|_{t=0} = \phi(x) \tag{11.3}$$

$(0 < \varepsilon \leq \varepsilon_0)$ for $0 \leq t \leq T$ can be written in the form

$$u(t, x) = w(t, x) + v(t, x', \rho/\varepsilon) + ... + \mathcal{O}(\varepsilon^2). \tag{11.4}$$

Here $w(t, x)$ is the solution of the limit (for $\varepsilon = 0$) boundary value problem

$$\frac{\partial w}{\partial t} = \Delta w - f(w) - g(x), \quad x \in \Omega, \tag{11.5}$$

$$w|_{\partial\Omega} = 0, \tag{11.6}$$

$$w|_{t=0} = \phi(x), \tag{11.7}$$

and $v(t, x', \rho/\varepsilon)$ is the boundary layer function which can be represented in the form

$$v(t, x', \rho/\varepsilon) = \varepsilon a(t, x') e^{-\lambda\rho/\varepsilon} \tag{11.8}$$

near the boundary $\partial\Omega \times [0, T]$. Here (x', ρ) are the local coordinates: $x' \in \partial\Omega$, ρ is the distance from the point x to the boundary $\partial\Omega$ along the normal to $\partial\Omega$. Under certain conditions on $f(u)$ the expansion (11.4) can be extended with the aid of two iteration processes (see Vishik & Lusternik [1], [2]) to terms of the orders $\mathcal{O}(\varepsilon^k)$, and the remainder is of order ε^{k+1} in the proper norm.

We shall study the asymptotics in ε of elements of the attractors $\mathcal{A}^\varepsilon$ corresponding to the evolution equations (11.1), (11.2).

For any element $u \in \mathcal{A}^\varepsilon$ the main term of asymptotics is constructed. We restrict ourselves to dimension $n = 3$ for brevity. Under certain conditions on $f(u)$ it is proved that for any function $u = u(x) \in \mathcal{A}^\varepsilon$ the asymptotic expansion

$$u = (u_0 + v_0 - \varepsilon c_0) + \varepsilon(u_1 + v_1 - \varepsilon c_1) + r \tag{11.9}$$

holds, where u_0 and u_1 are bounded in $H_3(\Omega)$ uniformly for ε, $0 < \varepsilon \le \varepsilon_0$, and for $u \in \mathcal{A}^\varepsilon$, v_0 and v_1 are boundary layer terms, which are constructed by u_0 and u_1, c_0 and c_1 are sufficiently smooth functions, which are bounded in proper norms uniformly for ε and for $u \in \mathcal{A}^\varepsilon$. The remainder term r satisfies the following estimate:

$$\|r\|_{H_3(\Omega)} + \varepsilon^{-1}\|r\|_{H_2(\Omega)} + \varepsilon^{-2}\|r\|_{H_1(\Omega)} \le M, \tag{11.10}$$

where M does not depend on ε and on $u \in \mathcal{A}^\varepsilon$. Thus it is proved that the boundary layer phenomenon which holds for solutions $u(t, x)$ of the problem (11.1) - (11.3) on any finite interval $[0, T]$ applies to the elements of the attractor $\mathcal{A}^\varepsilon$ which is the ω-limit set for $S_t(\varepsilon)B_0$ where B_0 is the absorbing set for the semigroup $\{S_t(\varepsilon)\}$.

It should be noted that the boundedness, uniform for ε, of attractors $\mathcal{A}^\varepsilon$ corresponding to the problem (11.1) - (11.2) (for $n = 3$) in $H_{3/2}(\Omega)$ was established by M. Skvortsov [1], [2]:

$$\|\mathcal{A}^\varepsilon\|_{3/2} \leq M, \tag{11.11}$$

M does not depend on ε, and the index $3/2$ cannot be increased. Moreover, it was proved that

$$\text{dist}_{H_{3/2-\delta}}(\mathcal{A}^\varepsilon, \mathcal{A}^0)_{\varepsilon \to 0} \to 0 \quad \forall \delta > 0, \tag{11.12}$$

where $\mathcal{A}^0$ is the attractor of the degenerate problem (11.5) - (11.6) and in (11.12) δ cannot be equal to zero. (There is given by M. Skvortsov [3] an example showing that the distance between the equilibrium points $z_\varepsilon(x)$ of $\{S_t(\varepsilon)\}$, which are the solutions of a linear ordinary differential equation, and the equilibrium point $z_0(x)$ of the limit semigroup $\{S_t(0)\}$ need not tend to zero as $\varepsilon \to 0$ in $H_{3/2}(\Omega)$. In this case the attractor $\mathcal{A}^\varepsilon$ consists of only one point $z_\varepsilon(x)$, therefore the assertion (11.12) is wrong when $\delta = 0$.)

§12 Construction of the asymptotic expansion

We give here the precise construction of the asymptotic expansion and the basic theorems. Assume for simplicity that $n = 3$ and Ω is the cylinder domain: $\Omega = T_x^2 \times \{0 < y < 1\}$ where T_x^2 is the two-dimensional torus whose points are denoted by x $(x = (x_1, x_2))$. It is not difficult to extend our results for any n and any domains Ω with sufficiently smooth boundary. Let $f(u)$ satisfy the following conditions:

$$f(u)u \geq -C, \tag{12.1}$$

$$f'(u) \geq -C, \quad |f''(u)| \leq C(|u| + 1), \tag{12.2}$$

$$|f^{(\ell)}(u)| \leq C, \quad 3 \leq \ell \leq k, \quad k \geq 7. \tag{12.3}$$

Assume also that

$$-C_1 \leq F(u) \equiv \int_0^u f(\xi)\,d\xi \leq C_2(f(u)u + u^2 + 1). \tag{12.4}$$

It is easy to establish by the methods described, for instance, in Babin & Vishik [1] or Lions [1], that there exists a unique $u(t)$ of (11.1) - (11.3) in corresponding functional spaces. Hence, to the problem (11.1)

- (11.3) corresponds the semigroup $\{S_t(\varepsilon)\}$, $0 < \varepsilon \leq \varepsilon_0$, $S_t(\varepsilon) : H \to H$ $\forall t > 0$, $H = L_2(\Omega)$, defined by $S_t(\varepsilon)\phi = u(t)$. It is easy to show (see Babin & Vishik [1]) that for any $\varepsilon > 0$ the semigroup $\{S_t(\varepsilon)\}$ has the (H, H)-attractor $\mathcal{A}^\varepsilon$, i.e. a compact set in H, which satisfies the invariance condition $(S_t(\varepsilon)\mathcal{A}^\varepsilon = \mathcal{A}^\varepsilon$ $\forall t \geq 0)$ and the attractive condition $(\mathrm{dist}_H(S_t(B), \mathcal{A}^\varepsilon) \to 0$ as $t \to +\infty$ for any bounded set B in $H)$. Moreover, as is shown in M. Skvortsov [1], [2], these attractors are bounded in $H_{3/2}(\Omega)$ uniformly for ε, $0 < \varepsilon \leq \varepsilon_0$ (see (11.11)). For $\varepsilon = 0$ the problem (11.1) - (11.3) reduces to the limit problem (11.5) - (11.7), in which only one boundary condition is given. This problem also has the (H, H)-attractor $\mathcal{A}^0$ and $\mathcal{A}^\varepsilon \to \mathcal{A}^0$ in $H_{3/2-\delta}(\Omega)$ $\forall \delta > 0$ as $\varepsilon \to 0$ (see (11.12)).

The key task is to find the main term of asymptotics on ε of elements $u \in \mathcal{A}^\varepsilon$. As will be shown below the expansion (11.9) is valid for $u \in \mathcal{A}^\varepsilon$ where $u_0 = u_0(x, y; \varepsilon, u)$, $u_1 = u_1(x, y; \varepsilon, u)$ are uniformly bounded in $H_3(\Omega)$ for ε, $0 < \varepsilon \leq \varepsilon_0$, v_0 and v_1 are the boundary layer functions, c_0 and c_1 are sufficiently smooth functions and $r(\varepsilon)$ is the remainder satisfying the estimates (11.10).

First we define more precisely each term in (11.9) and formulate the main theorem. Let us take any element $u = u(x, y) \in \mathcal{A}^\varepsilon$ ($x \in T^2$, $0 < y < 1$) and denote by $u(t) = u(t, x, y)$, $t \in \mathbf{R}$, the trajectory of $\{S_t(\varepsilon)\}$ passing through u when $t = 1$: $u(1) = u$ (more explicitly $u(1, x, y) = u(x, y)$ $\forall (x, y) \in \Omega$). The existence of such a trajectory $u(t)$ ($t \in \mathbf{R}$) follows from the properties of the attractor $\mathcal{A}^\varepsilon$ (see, for instance, Babin & Vishik [1]). Below we shall consider only the part of this trajectory which corresponds to the interval $0 \leq t \leq 1$ and $u(0) \in \mathcal{A}^\varepsilon$. Since $u(t)$ is the trajectory of the semigroup $\{S_t(\varepsilon)\}$ which lies on the attractor $\mathcal{A}^\varepsilon$, it satisfies the equation (11.1). Therefore we have

$$L_\varepsilon u(1) \equiv -\varepsilon^2 \Delta^2 u(1) + \Delta u(1) = \partial_t u(1) + f(u(1)) + g \equiv h, \qquad (12.5)$$

$h = h(x, y, \varepsilon)$. In order to obtain the first approximation of the function u, we consider the solution u_0 of the following boundary value problem:

$$\Delta u_0 = h \qquad (12.6)$$

where h is the right-hand side of (12.5),

$$u_0|_{\partial\Omega} = 0. \qquad (12.7)$$

Obviously this problem has a unique solution and as will be shown this solution $u_0 = u_0(x, y; \varepsilon, u)$ is bounded in $H_3(\Omega)$ uniformly for ε and $u \in \mathcal{A}^\varepsilon$:

$$\|u_0\|_{H_3(\Omega)} \leq M \qquad (12.8)$$

(and its derivatives by variables x_i are uniformly bounded in $H_{k+1,3}(\Omega)$). But this approximation u_0 for u is not valid in the neighbourhood of the boundary $\partial\Omega$, since the second boundary condition (11.2) for

$$\frac{\partial u_0}{\partial \nu}\,|_{\partial\Omega} \equiv \frac{\partial u_0}{\partial y}\,|_{y=0,y=1}$$

is not satisfied. A better approximation to u can be obtained by adding to u_0 a boundary layer term v_0 (see Vishik & Lusternik [1], [2]). In the neighbourhood of $\Gamma_0 = \partial\Omega \cap \{y = 0\}$ the function v_0 looks like

$$\tilde{v}_0(x,y;\varepsilon) = \varepsilon\tilde{c}_0(x)e^{-y/\varepsilon} \tag{12.9}$$

(compare with (11.8)). It should be noted that $\tilde{v}_0$ satisfies the differential equation

$$-\varepsilon^2\frac{\partial^4\tilde{v}_0}{\partial y^4} + \frac{\partial^2\tilde{v}_0}{\partial y^2} = 0$$

and $|\tilde{v}_0(x,y;\varepsilon)|$ exponentially decreases for increasing values of y and for decreasing values of ε ($\varepsilon > 0$). The function $\tilde{c}_0(x)$ is determined such that the second boundary condition (11.2) is satisfied:

$$\frac{\partial(u_0 + \tilde{v}_0)}{\partial y}\,|_{\Gamma_0} = 0,$$

i.e.

$$\frac{\partial\tilde{v}_0}{\partial y}\,|_{y=0} = -\tilde{c}_0(x) = \frac{\partial u_0}{\partial y}\,|_{y=0}. \tag{12.10}$$

As will be shown the function $\tilde{c}_0(x)$ satisfies the following estimate:

$$\|\tilde{c}_0(x)\|_{H_{(k+1)/2}(\partial\Omega)} \leq M, \tag{12.11}$$

where M is independent of ε and of $u \in \mathcal{A}^\varepsilon$. Similarly the function $\tilde{v}_0$ is determined in the neighbourhood of $\Gamma_1 = \partial\Omega \cap \{y = 1\}$.

In order to express the boundary layer term as a function defined in all the domain Ω, we multiply $\tilde{v}_0$ by an infinitely differentiable smoothing function $\chi(y)$:

$$\chi(y) = \begin{cases} 1 & \text{if } y \in [0,\eta] \cup [1-\eta, 1], \\ 0 & \text{if } y \in [2\eta, 1-2\eta], \end{cases} \tag{12.12}$$

where η is sufficiently small. In this way we define the second term in (11.9):

$$v_0 = \chi(y)\tilde{v}_0(x,y;\varepsilon).$$

Now the function $u_0 + v_0$ satisfies (by (12.10)) the second boundary condition (11.2), but does not satisfy the first one, since

$$(u_0 + v_0)|_{\Gamma_0} = \tilde{v}_0|_{\Gamma_0} = \varepsilon \tilde{c}_0(x)$$

where $\tilde{c}_0(x)$ is defined in (12.10). A similar relation holds on Γ_1 when we replace $\tilde{c}_0(x)$ by $\tilde{c}_0^1(x)$. Let $c_0(x, y)$ denote the function which is equal to $\chi(y)\tilde{c}_0(x)$ in a 2η-neighbourhood of Γ_0, equal to $\chi(y)\tilde{c}_0^1(x)$ in a 2η-neighbourhood of Γ_1 and equal to zero for $2\eta < y < 1 - 2\eta$.

Thus the function

$$u^0 \equiv u_0 + v_0 - \varepsilon c_0$$

satisfies the boundary condition (11.2) and roughly satisfies the equation (12.5) up to order ε. Therefore it may be considered as the first approximation for u. However, it turns out that it is not enough for obtaining the estimates (11.10) for the remainder $r(\varepsilon)$. Therefore we add to the function u^0 the following terms:

$$\varepsilon u^1 \equiv \varepsilon(u_1 + v_1 - \varepsilon c_1),$$

so that the function $u^0 + \varepsilon u^1$ satisfies the equation (12.5) more exactly. Substituting this function into (12.5), using (12.6) and equating to zero the coefficient of ε (leaving out of account the boundary layer) we get the differential equation

$$\Delta u_1 = \Delta c_0. \tag{12.13}$$

We supply this equation with the first of the boundary conditions (11.2):

$$u_1|_{\partial\Omega} = 0.$$

Hence, we define the function $u_1 = u_1(x, y; \varepsilon, u)$. It follows from the equation (12.13) that for u_1, as will be shown, the estimate similar to (12.11) is valid:

$$\|u_1(x, y; \varepsilon, u)\|_{H_{(k+1)/2}(\Omega)} \le M \tag{12.14}$$

and M is independent of ε and of $u \in \mathcal{A}^\varepsilon$. In order to satisfy the second of the boundary conditions (11.2), we add to the function $u^0 + \varepsilon u_1$ a boundary layer function which looks similar to (12.9) (in the neighbourhood of Γ_1):

$$\varepsilon \tilde{v}_1 = \varepsilon^2 \tilde{c}_1(x) e^{-y/\varepsilon}.$$

Similarly to (12.10) we obtain

$$\frac{\partial \tilde{v}_1}{\partial y}\Big|_{y=0} = -\tilde{c}_1(x) = \frac{\partial u_1}{\partial y}\Big|_{y=0}. \tag{12.15}$$

Then, as above, we set $v_1 = \chi(y)\tilde{v}_1$ and to satisfy the first boundary condition (11.2) add the function $-\varepsilon^2 c_1(x, y)$. As will be shown in subsequent sections the following estimate holds:

$$\varepsilon\|c_1\|_{H_{(5k-2)/8}(\Omega)} + \|c_1\|_{H_{(k-2)/2}(\Omega)} \leq M, \qquad (12.16)$$

where M is independent of ε and of $u \in \mathcal{A}^\varepsilon$. Thus all terms in the asymptotic expansion (11.9) are determined.

Now we formulate our main result.

Theorem 12.1

The remainder $r(x, y; \varepsilon)$ satisfies the estimate (11.10) in which M is independent of $u \in \mathcal{A}^\varepsilon$ and of ε, $0 < \varepsilon \leq \varepsilon_0$.

The proof will be given in the next section. We shall now establish some corollaries following from this theorem.

Theorem 12.2.

Let $\Omega_\gamma = \{(x, y) \in \Omega \mid \gamma < y < 1 - \gamma\}$. Then for any $u \in \mathcal{A}^\varepsilon$

$$\|u\|_{H_3(\Omega_\gamma)} \leq M, \qquad (12.17)$$

where $M = M(\gamma)$ is independent of $u \in \mathcal{A}^\varepsilon$ and of ε, $0 < \varepsilon \leq \varepsilon_0$.

Proof It follows from the expansion (11.9) that

$$\begin{aligned}
\|u\|_{H_3(\Omega_\gamma)} \leq &\|u_0\|_{H_3(\Omega_\gamma)} + \|v_0\|_{H_3(\Omega_\gamma)} + \varepsilon\|c_0\|_{H_3(\Omega_\gamma)} \\
&+ \varepsilon(\|u_1\|_{H_3(\Omega_\gamma)} + \|v_1\|_{H_3(\Omega_\gamma)} + \varepsilon\|c_1\|_{H_3(\Omega_\gamma)}) \qquad (12.18) \\
&+ \|r(\varepsilon)\|_{H_3(\Omega_\gamma)}.
\end{aligned}$$

The first, third, fourth and sixth terms in the right-hand side are bounded according to (12.8), (12.11), (12.14) and (12.16). From (12.9) and (12.11) we have

$$\|v_0\|_{H_3(\Omega_\gamma)} \leq M\varepsilon\|e^{-y/\varepsilon}\|_{H_3((\gamma, 1-\gamma))} \leq M_1.$$

Analogously it follows from the boundedness of $\varepsilon\|c_1\|_{H_3(\Omega)}$ that

$$\|v_1\|_{H_3(\Omega_\gamma)} \leq M.$$

We finally note that Theorem 12.1 implies that the last term in (12.18) is also bounded.

Remark 12.1.

It follows from Theorem 12.2 that elements of the attractors $\mathcal{A}^\varepsilon$ are bounded uniformly for $u \in \mathcal{A}^\varepsilon$ and for ε in $H_3(\Omega_\gamma)$, where $\Omega_\gamma \subseteq \Omega$ is any internal subdomain; meanwhile we have already seen that in the whole domain Ω these elements are bounded uniformly only in $H_{3/2}(\Omega)$ (see (11.11)), and the index 3/2 cannot be increased.

Remark 12.2.

It is easily shown (see Babin & Vishik [1], M. Skvortsov [2]), that for $\varepsilon = 0$

$$\|\mathcal{A}^0\|_{H_3(\Omega)} \le M.$$

Theorem 12.3.

For any $\gamma > 0$, $\delta > 0$

$$\mathrm{dist}_{H_{3-\delta}(\Omega_\gamma)}(\mathcal{A}^\varepsilon, \mathcal{A}^0)_{\varepsilon \to 0} \to 0. \tag{12.19}$$

Proof   First we remark that

$$\mathrm{dist}_H(\mathcal{A}^\varepsilon, \mathcal{A}^0)_{\varepsilon \to 0} \to 0 \tag{12.20}$$

(where $H = L_2(\Omega)$). Indeed, let $\{S_t(\varepsilon)\}$ be the semigroup, corresponding to the boundary value problem (11.1) - (11.3) for $0 < \varepsilon \le \varepsilon_0$ and to the problem (11.5) - (11.7) for $\varepsilon = 0$. Then on the set $X = B_0 \times \Lambda_0 \subseteq H \times \Lambda_0$ where $\Lambda_0 = [0, \varepsilon_0]$, the semigroup $\{S_t = (S_t(\varepsilon), \varepsilon)\}$ satisfies all the conditions of Theorem 1 in the book Babin & Vishik [1] on page 226: this semigroup is continuous on X in $H \times \Lambda_0$ and $\{S_t(\varepsilon)\}$ has a set B_0 which is H-absorbing uniformly for $\varepsilon \in \Lambda_0$ and compact in H (as was shown by M. Skvortsov [3]). Hence, according to Proposition 3 in Babin & Vishik [1] on page 227, the assertion (12.20) is valid. Since the attractors $\mathcal{A}^\varepsilon$ are bounded uniformly in $H_3(\Omega_\gamma)$ as has been proved in Theorem 12.2, then there exists a uniformly absorbing set B_1 for $\varepsilon \in \Lambda_0$ and compact in $H_{3-\delta}(\Omega_\gamma)$ (it consists of the functions $u \in \mathcal{A}^\varepsilon$ which we consider only on Ω_γ), $\delta > 0$. Then the assertion (12.19) follows from Remark 1 in the book Babin & Vishik [1] on page 228.

Remark 12.3.

It was obtained by M. Skvortsov [2], [3] that

$$\mathrm{dist}_{H_{3/2-\delta}(\Omega)}(\mathcal{A}^\varepsilon, \mathcal{A}^0)_{\varepsilon \to 0} \to 0 \quad \forall \delta > 0.$$

It follows from Theorem 12.3 that $H_{3/2-\delta}(\Omega)$ may be replaced by

$$H_{3-\delta}(\Omega_\gamma) \quad (\delta > 0, \gamma > 0).$$

This fact shows that the functions $u \in \mathcal{A}^\varepsilon$ have boundary layer properties near $\partial\Omega$.

Remark 12.4.

Taking into account that the attractors $\mathcal{A}^\varepsilon$ are bounded in $H_{3/2}(\Omega)$ uniformly for ε (see (11.11), we obtain that any set $\{u_\varepsilon \mid u_\varepsilon \in \mathcal{A}^\varepsilon, \varepsilon \to 0\}$ is compact in H and therefore we can choose a convergent sub-sequence from it:

$$u_{\varepsilon_m} \to w \quad \text{in } H \text{ as } \varepsilon_m \to 0. \tag{12.21}$$

But $\mathcal{A}^{\varepsilon_m} \to \mathcal{A}^0$ as $\varepsilon_m \to 0$, hence $w = \lim_{\varepsilon_m \to 0} u_{\varepsilon_m}$ is an element of $\mathcal{A}^0$.

Theorem 12.4.

Let (12.21) be valid. Then
(1)
$$u_{\varepsilon_m} \to w \text{ in } H_{3-\delta}(\Omega_\gamma) \text{ as } \varepsilon_m \to 0 \tag{12.22}$$

for any $\delta > 0$, $\gamma > 0$;
(2) in the asymptotic expansion for u_{ε_m}

$$u_{\varepsilon_m} = (u_{0m}+v_{0m}-\varepsilon_m c_{0m})+\varepsilon_m(u_{1m}+v_{1m}-\varepsilon_m c_{1m})+r(\varepsilon_m) \tag{12.23}$$

the following assertions for u_{0m} and c_{0m} hold:

$$u_{0m} \to w \text{ in } H_{3-\delta}(\Omega), \quad \forall \delta > 0, \tag{12.24}$$

$$c_{0m} \to \frac{\partial w}{\partial y}\mid_{y=0} \text{ in } H_{(k+1)/2-\delta_1}(\Omega), \qquad \delta_1 > 0, \tag{12.25}$$

near Γ_0 and the analogous assertion for c_{0m} is valid in a neighbourhood of Γ_1.

Proof   Theorem 12.2 implies that the set $\{u_{\varepsilon_m}\}$ is compact in

$$H_{3-\delta}(\Omega_\gamma)$$

when $\delta > 0$. Then assertion (12.22) follows from (12.21).

Taking into account the expression (12.9) for the boundary layer term v_{0m} and using the estimate (12.11), we obtain that

$$\|v_{0m}\|_{H_{3/2-\delta}(\Omega)} \to 0, \quad \varepsilon_m\|c_{0m}\|_{H_{(k+1)/2}(\Omega)} \to 0 \text{ as } \varepsilon_m \to 0.$$

Similarly it follows from (12.16) that

$$\varepsilon_m \|v_{1m}\|_{H_{5/2-\delta}(\Omega)} \to 0, \quad \varepsilon_m^2 \|c_{1m}\|_{H_{(5k-2)/8}(\Omega)} \to 0 \text{ as } \varepsilon_m \to 0.$$

From the estimate (12.14) we have

$$\varepsilon_m \|u_{1m}\|_{H_{(k+1)/2}(\Omega)} \to 0 \text{ as } \varepsilon_m \to 0$$

and inequality (11.10) implies

$$\|r(\varepsilon_m)\|_{H_{3-\delta}(\Omega)} \to 0 \text{ as } \varepsilon_m \to 0.$$

Hence we see from (12.21) and (12.23) that

$$\lim_{\varepsilon_m \to 0} u_{0m} = \lim_{\varepsilon_m \to 0} u_{\varepsilon_m} = w,$$

where the limits are taken, for instance, in the metric of $H_{3/2-\delta}$. Since the sequence $\{u_{0m}\}$ is uniformly bounded in $H_3(\Omega)$, it is compact in $H_{3-\delta}(\Omega)$ $\forall \delta > 0$ and we can choose a sub-sequence converging to the same limit w in $H_{3-\delta}(\Omega)$. Since this assertion is valid for any sub-sequence, (12.24) is also valid.

It follows from (12.24) now that

$$\frac{\partial u_{0m}}{\partial y}\Big|_{y=0} \to \frac{\partial w}{\partial y}\Big|_{y=0}, \tag{12.26}$$

for instance, in the metric of $H_{3/2-\delta}(\Omega)$. Taking into account that in a neighbourhood of Γ_0

$$c_{0m} = \frac{\partial u_{0m}}{\partial y}\Big|_{y=0}$$

(see (12.10)) and the functions c_{0m} are uniformly bounded in $H_{(k+1)/2}(\Omega)$ (see (12.11)) we deduce (12.25) from (12.26). Analogously we consider the properties of c_{0m} in a neighbourhood of Γ_1.

Remark 12.5.

The assertions (12.24) and (12.25) show that the expansion (11.9) and the asymptotic expansion of the solution of stationary differential equation by the procedure of Lusternik & Vishik are close in some way.

§13   Proof of Theorem 12.1

We turn now to the proof of the main Theorem 12.1. First we shall formulate some estimates for the trajectories $u(t)$ of the problem (11.1) - (11.3). We denote norms in Sobolev spaces $H_s(\Omega)$ by $||\cdot||_s$, in $H = L_2(\Omega)$ by $||\cdot||$.

Proposition 13.1.
Let the function $f(u)$ satisfy condition (12.1) and $g \in L_2(\Omega)$. Then the following estimate for the solution $u(t)$ of the problem (11.1) - (11.3) holds:

$$||u(t)||^2 + \int_0^t \left(||u(\tau)||_1^2 + \varepsilon^2 ||u(\tau)||_2^2 \right) \, \mathrm{d}\tau \le C. \qquad (13.1)$$

If, furthermore, condition (12.4) is satisfied, then the following estimate holds:

$$\int_0^t \tau ||u_t||^2 \, \mathrm{d}\tau + \varepsilon^2 t ||u(t)||_2^2 + t ||u(t)||_1^2 \le C. \qquad (13.2)$$

In the inequalities (13.1) and (13.2) $t \in [0, T]$ and the constants C depend only on $||u(0)||$ and T.

Proposition 13.2.
Let the function $f(u)$ satisfy (12.1), (12.2) and (12.4) and $g \in L_2(\Omega)$. Then the following estimate holds:

$$\int_0^t \tau^3 ||u_{tt}||^2 \, \mathrm{d}\tau + \varepsilon^2 t^3 ||u_t||_2^2 + t^3 ||u_t||_1^2 \le C, \qquad (13.3)$$

where $t \in [0, T]$ and C depends only on $||u(0)||$ and T.

Proposition 13.3.
Let $f(u)$ satisfy conditions (12.1) - (12.4) and the function $g(x)$ and its derivatives $\partial^j g / \partial x_i^j$ ($i = 1, 2$, $j = 1, 2, ..., k - 2$) belong to $H_2(\Omega)$. Denote by $p_\ell(t)$ the derivative $\partial^\ell u(x, y, t) / \partial x_i^\ell$, $i = 1$ or 2. Then the following estimates, uniform for ε, $0 \le \varepsilon \le \varepsilon_0$, are valid:

$$\varepsilon^{5/2} ||p_\ell(1)||_4 + \varepsilon^{3/2} ||p_\ell(1)||_3 + \varepsilon^{1/2} ||p_\ell(1)||_2 + ||p_\ell(1)||_{3/2} \le C, \qquad (13.4)$$

$$\varepsilon ||\partial_t p_\ell(1)||_2 + ||\partial_t p_\ell(1)||_1 \le C, \qquad (13.5)$$

where $\ell = 1, 2, ..., k - 2$ and the constants C depend only on $||u(0)||$.

Propositions 13.1, 13.2 and 13.3 will be proved in §14.

Denote by $W_2^{(\ell_1,\ell_2,\ell_3)}(\Omega)$ a non-isotropic functional space (see Besov, Ilyin & Nikolsky [1]) with the norm

$$\|u\|_{W_2^{\underline{\ell}}} = \|u\| + \sum_{i=1}^{3} \|D_i^{l_i} u\|, \quad \underline{\ell} = (\ell_1, \ell_2, \ell_3),$$

where $D_1 = \partial/\partial x_1$, $D_2 = \partial/\partial x_2$, $D_3 = \partial/\partial y$. The following embedding theorems are valid (Nikolsky [1], Besov, Ilyin & Nikolsky [1]).

(1) If a function $u \in W_2^{\underline{\ell}}(\mathbf{R}^n)$ then its partial derivative

$$D^{\underline{k}} u \in W_2^{\psi \underline{\ell}}(\mathbf{R}^n), \tag{13.6}$$

where

$$\psi = 1 - \sum_{j=1}^{n} \frac{k_j}{\ell_j} > 0.$$

(2)

$$W_2^{\underline{\ell}}(\mathbf{R}^n) \hookrightarrow W_2^{\psi \underline{\ell}^m}(\mathbf{R}^m), \tag{13.7}$$

where

$$\underline{\ell} = (\ell_1, ..., \ell_n), \quad \underline{\ell}^m = (\ell_1, ..., \ell_m, 0, ..., 0),$$

$$\psi = 1 - \frac{1}{2} \sum_{j=m+1}^{n} \frac{1}{\ell_j} > 0, \quad m < n.$$

We shall apply these theorems to the terms in the asymptotic expansion (11.9).

Proposition 13.4.
Under the assumptions of Proposition 13.3 the following estimates uniform for ε for the functions u_0, u_1, c_0, c_1 in (11.9) hold:

$$\varepsilon \|u_0\|_{W_2^{(k+2,4)}} + \|u_0\|_{W_2^{(k+1,3)}} \le C; \tag{13.8}$$

here the first indices show the orders of the derivatives with respect to x_1 and x_2 and the second the orders of derivatives with respect to y,

$$\varepsilon \|u_1\|_{5(k+2)/8} + \|u_1\|_{(k+1)/2} \le C, \tag{13.9}$$

$$\varepsilon \|c_0\|_{W_2^{(5(k+2)/8, \infty)}} + \|c_0\|_{W_2^{((k+1)/2, \infty)}} \le C, \tag{13.10}$$

$$\varepsilon \|c_1\|_{W_2^{(5k-2)/8, \infty)}} + \|c_1\|_{W_2^{((k-2)/2, \infty)}} \le C. \tag{13.11}$$

Proof   We consider the boundary value problem (12.6) - (12.7) for u_0:

$$\Delta u_0 = \partial_t u(1) + f(u(1)) + g \equiv h(\varepsilon), \qquad (12.6)$$

$$u|_{\partial\Omega} = 0. \qquad (12.7)$$

The inequalities (13.4) and (13.5) imply the following estimate for the right-hand side $h(\varepsilon)$:

$$\varepsilon ||h||_{W_2^{(k,2)}} + ||h||_{W_2^{(k-1,1)}} \le C. \qquad (13.12)$$

Indeed, this estimate for the function $\partial_t u(1)$ follows immediately from (13.5) where $\ell = k - 2$. Further, since

$$D_{x_i}^\ell f(u) = f^{(\ell)}(u)p_1^\ell + a_{(\ell-1)}f^{\ell-1}(u)p_1^{\ell-2}p_2 + \dots + f'(u)p_\ell,$$

the conditions (12.2), (12.3) on the derivatives of $f^{(m)}(u)$ and the estimates (13.4) imply the existence (13.12) for $f(u(1))$. We use here the Hölder inequality for estimating of derivatives

$$D_{x_i}^k f(u(1)) \quad \text{and} \quad D_{x_i}^{k-1} f(u(1))$$

and also the Sobolev embedding theorems

$$\begin{aligned} H_{3/2}(\Omega) &\subseteq L_p(\Omega) \quad \forall p, \\ H_{1/2}(\Omega) &\subseteq L_3(\Omega), \end{aligned} \qquad (13.13)$$

which are valid for $n = 3$. It follows from the estimate (13.12) that the function u_0, as the solution of the elliptic boundary value problem (12.6) - (12.7), satisfies inequality (13.8).

We now consider the functions $c_0(x,y)$ and $c_1(x,y)$. Observe at first that their infinite differentiability with respect to y follows from their construction. Since $c_0(x,y) \equiv \tilde{c}_0(x)$ and $c_1(x,y) \equiv \tilde{c}_1(x)$ in a neighbourhood of the boundary Γ_0, where $\tilde{c}_0(x)$ and $\tilde{c}_1(x)$ have been defined in (12.10) and (12.15) respectively, c_0 and c_1 are defined in a neighbourhood of Γ_1 analogously and these functions are equal to zero in the internal subdomain $\Omega_{2\eta}$, it is enough to prove the inequalities (13.10) and (13.11) only for $\tilde{c}_0(x)$ and $\tilde{c}_1(x)$.

We shall estimate

$$\tilde{c}_0(x) \equiv \left(\frac{\partial u_0}{\partial y}\right)|_{y=0}$$

by applying (13.6) to the function on u_0. Putting in (13.6) $\underline{k} = (0,0,1)$, $\underline{\ell}_1 = (k+2, k+2, 4)$, $\psi_1 = 3/4$, $\underline{\ell}_2 = (k+1, k+1, 3)$, $\psi_2 = 2/3$, we get, using the estimates (13.8) for u_0:

$$\varepsilon\left\|\frac{\partial u_0}{\partial y}\right\|_{W_2^{(\frac{3}{4}(k+2),3)}} + \left\|\frac{\partial u_0}{\partial y}\right\|_{W_2^{(\frac{2}{3}(k+1),2)}} \leq C. \qquad (13.14)$$

We apply now the embedding theorem (13.7) to the function $\partial u_0/\partial y$. Putting in (13.7)

$$\ell_1 = \left(\frac{3}{4}(k+2), \frac{3}{4}(k+2), 3\right), \quad \psi_1 = 1 - \frac{1}{2}\cdot\frac{1}{3} = \frac{5}{6},$$

$$\psi_1\ell_1^m = \left(\frac{5}{8}(k+2), \frac{5}{8}(k+2), 0\right),$$

$$\ell_2 = \left(\frac{2}{3}(k+1), \frac{2}{3}(k+1), 2\right), \quad \psi_2 = 1 - \frac{1}{2}\cdot\frac{1}{2} = \frac{3}{4},$$

$$\psi_2\ell_2^m = \left(\frac{1}{2}(k+1), \frac{1}{2}(k+1), 0\right),$$

we get from (13.14) the following estimate for $\tilde{c}_0(x)$:

$$\varepsilon\|\tilde{c}_0(x)\|_{H_{5(k+2)/8}(\Gamma_0)} + \|\tilde{c}_0(x)\|_{H_{(k+1)/2}(\Gamma_0)} \leq C.$$

From this inequality the estimate (13.10) follows.

The equation (12.13) and the inequality (13.10) imply the estimate (13.9) for u_1, from which in its turn the estimate (13.11) for $c_1(x,y)$ follows (since

$$c_1(x,y) \equiv \tilde{c}_1(x) = \left(\frac{\partial u_1}{\partial y}\right)\Big|_{y=0}$$

in a neighbourhood of Γ_0 and analogously in a neighbourhood of Γ_1).

Remark 13.1.

If $k = 7$ in (12.3), then from the estimate (13.11), which looks in this case like

$$\varepsilon\|c_1\|_{33/8} + \|c_1\|_{5/2} \leq C,$$

and from the interpolation inequality

$$\|u\|_{W_p^s} \leq C\|u\|_{W_p^{s_0}}^{1-\theta}\|u\|_{W_p^{s_1}}^{\theta},$$

$s = (1 - \theta)s_0 + \theta s_1$, in which we put $p = 2$, $s = 4$, $s_0 = 5/2$, $s_1 = 33/8$, $\theta = 12/13$, we have

$$\varepsilon^{12/13}||c_1||_4 + ||c_1||_{5/2} \leq C_1. \qquad (13.15)$$

Proof of Theorem 12.1   We remark that, since the functions u, $u^0 = u_0 + v_0 - \varepsilon c_0$ and $u^1 = u_1 + v_1 - \varepsilon c_1$ satisfy the boundary conditions (11.2), then the remainder $r(\varepsilon)$ also satisfies homogeneous boundary conditions:

$$r|_{\partial\Omega} = \frac{\partial r}{\partial y}\Big|_{\partial\Omega} = 0.$$

Consider the expression

$$
\begin{aligned}
L_\varepsilon r &\equiv -\varepsilon^2\Delta^2 r + \Delta r = L_\varepsilon(u - (u_0 + v_0 - \varepsilon c_0) - \varepsilon(u_1 + v_1 - \varepsilon c_1)) \\
&= h + \varepsilon^2\Delta^2 u_0 - \Delta u_0 - L_\varepsilon v_0 + \varepsilon(\varepsilon^2\Delta^2 u_1 - \Delta u_1) - \varepsilon L_\varepsilon v_1 \\
&\quad - \varepsilon(\varepsilon^2\Delta^2 c_0 - \Delta c_0) - \varepsilon^2(\varepsilon^2\Delta^2 c_1 - \Delta c_1) \\
&= \varepsilon^2\Delta^2 u_0 - \varepsilon^3\Delta^2 c_0 - \varepsilon^2(\varepsilon^2\Delta^2 c_1 - \Delta c_1) - L_\varepsilon v_0 - \varepsilon L_\varepsilon v_1 + \varepsilon^3\Delta^2 u_1
\end{aligned}
$$
$$(13.16)$$

(here we use equations (12.6), (12.13); the function h is defined in (12.5)).
    Multiplying $L_\varepsilon r$ by r in $L_2(\Omega)$ we have

$$|(L_\varepsilon r, r)| = \varepsilon^2(\Delta r, \Delta r) + ||r||_1^2. \qquad (13.17)$$

Substitute the expression for $L_\varepsilon r$ in the right-hand side of (13.16) into (13.17) and estimate inner products of each term:

$$
\begin{aligned}
|\varepsilon^2(\Delta^2 u_0, r)| &= |\varepsilon^2(\nabla(\Delta u_0), \nabla r)| \\
&\leq \frac{C\varepsilon^4}{\mu}||u_0||_3^2 + \mu||r||_1^2 \leq C_1\varepsilon^4 + \mu||r||_1^2
\end{aligned}
\qquad (13.18)
$$

(using (12.8));

$$
\begin{aligned}
|\varepsilon^3(\Delta^2 u_1, r)| &\leq C_1\varepsilon^6 + \mu||r||^2 \quad \text{(using (12.14), } k \geq 7\text{);} & (13.19) \\
|\varepsilon^3(\Delta^2 c_0, r)| &\leq C_1\varepsilon^6 + \mu||r||^2 \quad \text{(using (13.10));} & (13.20) \\
|\varepsilon^4(\Delta^2 c_1, r)| &\leq C_2\varepsilon^6 + \mu||r||^2 \quad \text{(using (13.11));} & (13.21) \\
|\varepsilon^2(\Delta c_1, r)| &\leq C_2\varepsilon^4 + \mu||r||^2 \quad \text{(using (13.11));} & (13.22)
\end{aligned}
$$

Taking into account that $v_0 = \varepsilon \tilde{c}_0(x)\chi(y)e^{-y/\varepsilon}$ (in a neighbourhood of Γ_0), we obtain

$$
\begin{aligned}
|(L_\varepsilon v_0, r)| &= |(\Delta(\varepsilon^2 \Delta v_0 - v_0), r)| \\
&\leq |(\varepsilon^2 \partial^2_{yy} v_0 - v_0, \Delta r)| + \varepsilon^2 |(\Delta_x v_0, \Delta r)| \\
&\leq \varepsilon^3 |((\chi'' - 2\varepsilon^{-1}\chi')\tilde{c}_0 e^{-y/\varepsilon}, \Delta r)| + \varepsilon^3 |(\chi e^{-y/\varepsilon}\Delta_x \tilde{c}_0, \Delta r)| \\
&\leq C_3 \varepsilon^4 + \mu_1 \varepsilon^2 \|\Delta r\|^2.
\end{aligned}
\tag{13.23}
$$

We have used

$$
|\chi^{(k)} e^{-y/\varepsilon}| \leq C_{k,N} \varepsilon^N \quad \forall k, N > 0
$$

(because $\chi^{(k)}(y) \equiv 0$ for $y \in [0, \eta] \cup [1 - \eta, \eta]$, and the estimate (12.11). Analogous considerations are valid for the function v_0 in a neighbourhood of Γ_1. In exactly the same way we can estimate the term with $L_\varepsilon v_1$:

$$
\varepsilon |(L_\varepsilon v_1, r)| \leq C_3 \varepsilon^4 + \mu_1 \varepsilon^2 \|\Delta r\|^2
\tag{13.24}
$$

(using (13.15)).

It follows from expressions (13.16) and (13.17) for $L_\varepsilon r$ and from the estimates (13.18) - (13.24), that

$$
\varepsilon^2 \|\Delta r\|^2 + \|r\|_1^2 = |(L_\varepsilon r, r)| \leq C_4 \varepsilon^4 + \mu_1 \varepsilon^2 \|\Delta r\|^2 + \mu \|r\|_1^2.
$$

From this estimate, choosing μ and μ_1 sufficiently small, we obtain

$$
\|r\|_2 = \|\Delta r\| \leq C\varepsilon, \quad \|r\|_1 \leq C\varepsilon^2.
\tag{13.25}
$$

Now we shall prove that

$$
\|L_\varepsilon r\| = \| - \varepsilon^2 \Delta^2 r + \Delta r\| \leq C\varepsilon.
\tag{13.26}
$$

Indeed, from (12.8), (12.14), (13.10) and (13.11) we deduce that norms in H of five terms in the right-hand side of (13.16) are of order ε:

$$
\varepsilon^2 \|\Delta^2 u_0\| + \varepsilon^3 \|\Delta^2 u_1\| + \varepsilon^3 \|\Delta^2 c_0\| + \varepsilon^2 \|\Delta c_1\| + \varepsilon^4 \|\Delta^2 c_1\| \leq C\varepsilon.
$$

Further, similarly to (13.23) we obtain

$$
\begin{aligned}
-L_\varepsilon v_0 &\equiv \Delta(\varepsilon^2 \Delta v_0 - v_0) = \Delta(\varepsilon^2 \Delta_x v_0 + \varepsilon^2 \partial^2_{yy} v_0 - v_0) \\
&= \varepsilon^3 \Delta(\Delta_x \tilde{c}_0 e^{-y/\varepsilon}\chi) + \varepsilon^2 \Delta_x \tilde{c}_0 e^{-y/\varepsilon}(-2\chi' + \varepsilon\chi'') \\
&\quad + \varepsilon^2 \tilde{c}_0 \partial^2_{yy}\left[e^{-y/\varepsilon}(-2\chi' + \varepsilon\chi'')\right].
\end{aligned}
$$

Hence from (12.11) we have

$$||L_\varepsilon v_0|| \leq C\varepsilon.$$

In a similar manner, using (13.15), we deduce

$$||L_\varepsilon v_1|| \leq C\varepsilon,$$

and (13.26) is therefore valid. It follows from (13.26) and (13.25) that

$$||\Delta^2 r|| \leq C\varepsilon^{-1}.$$

Combining this inequality and the first of the inequalities (13.25) and using the interpolation inequality

$$\varepsilon^k ||u||_k \leq C_{k,\ell}(\varepsilon^\ell ||u||_\ell + ||u||), \quad 0 < k < \ell, \qquad (13.27)$$

we obtain that the remainder $r(\varepsilon)$ is uniformly bounded in $H_3(\Omega)$:

$$||r||_3 \leq C.$$

Combining this result with the estimates (13.25), we obtain the assertion of Theorem 12.1.

Remark 13.2.

By differentiating (13.16) with respect to x_i ℓ times and using non-isotropic estimates (13.8) - (13.11), in exactly the same way as the inequality (11.10) has been proved, we obtain the following estimate, uniform for ε:

$$||r||_{3,\ell} + \varepsilon^{-1}||r||_{2,\ell} + \varepsilon^{-2}||r||_{1,\ell} \leq M,$$

where $||r||_{m,\ell} = ||r||_{W_2^{(m+\ell,m)}(\Omega)}$ and ℓ satisfies the following condition:

$$\frac{5}{8}k - \frac{17}{4} - \ell \geq 0.$$

§14 Proof of Propositions 13.1 - 13.3

We shall now prove Propositions 13.1, 13.2 and 13.3.

Proof of Proposition 13.1   Multiplying both sides of equation (11.1) by u and integrating on Ω, we obtain

$$\frac{1}{2}\partial_t\|u(t)\|^2 + \varepsilon^2\|\Delta u\|^2 + \|\nabla u\|^2 + \int f(u)u\,dx \le \mu\|u\|^2 + C\|g\|^2. \quad (14.1)$$

Choosing μ sufficiently small, using (12.1) and the Friedrichs inequality and integrating (14.1) with respect to t, we obtain (13.1).

To deduce (13.2) we multiply equation (11.1) by $t\partial_t u$ in $L_2(\Omega)$. Taking into account that

$$tf(u)u_t = \partial_t(tF(u)) - F(u)$$

we deduce

$$t\|u_t\|^2 + \varepsilon^2\partial_t(t\|\Delta u\|^2) + \partial_t(t\|\nabla u\|^2) + \partial_t\left(\int tF(u)\,dx\right)$$
$$\le \varepsilon^2\|\Delta u\|^2 + \|\nabla u\|^2 + \int F(u)\,dx + C\|g\|^2. \quad (14.2)$$

Then we integrate both sides of the inequality (14.2) with respect to t. Estimating the right-hand side obtained by means of (12.4), (13.1) and (14.1), we get the estimate (13.2).

Proof of Proposition 13.2   Let us differentiate both sides of equation (11.1) with respect to t:

$$u_{tt} = -\varepsilon^2\Delta^2 u_t + \Delta u_t - f'(u)u_t. \quad (14.3)$$

Multiplying (14.3) by $t^2 u_t$ in $L_2(\Omega)$ and using the first of the conditions (12.2) we have

$$\partial_t(t^2\|u_t\|^2) + \varepsilon^2 t^2\|\Delta u_t\|^2 + t^2\|\nabla u_t\|^2 \le 2t\|u_t\|^2 + Ct^2\|u_t\|^2.$$

Integrating this inequality with respect to t and using the estimate (13.2), we deduce

$$t^2\|u_t\|^2 + \int_0^t (\varepsilon^2\tau^2\|\Delta u_t\|^2 + \tau^2\|\nabla u_t\|^2)\,d\tau \le C, \quad (14.4)$$

$C = C(t)$. Then multiplying both sides of (14.3) by $t^3 u_{tt}$ in L_2 we obtain

$$t^3\|u_{tt}\|^2 + \varepsilon^2\partial_t(t^3\|\Delta u_t\|^2) + \partial_t(t^3\|\nabla u_t\|^2)$$
$$\le C(\varepsilon^2 t^2\|\Delta u_t\|^2 + t^2\|\nabla u_t\|^2 + t^3\|f'(u)\|_{L_3}\|u_t\|_{L_6}\|u_{tt}\|)$$
$$\le C(\varepsilon^2 t^2\|\Delta u_t\|^2 + t^2\|\nabla u_t\|^2 + 1) + \frac{1}{2}t^3\|u_{tt}\|^2$$

$(||f'(u)||_{L_3} \leq C_2(1 + ||u||_1^2) \leq C_3$; see (14.11)). (We have used (12.2), (11.11), the Hölder inequality and Sobolev embedding theorems). Integrating with respect to t and using (14.4), we obtain the estimate (13.3).

The following lemma (see M. Skvortsov [2], [3]) plays a basic role in the proof of Proposition 13.3.

Lemma 14.1.
Let $z(\varepsilon)$ be the solution of the stationary boundary value problem

$$L_\varepsilon z \equiv -\varepsilon^2 \Delta^2 z + \Delta z = h(\varepsilon) \ \text{in } \Omega \Subset \mathbf{R}^n, \tag{14.5}$$

$$z|_{\partial\Omega} = \frac{\partial z}{\partial \nu}\Big|_{\partial\Omega} = 0, \tag{14.6}$$

$\partial\Omega$ a sufficiently smooth manifold.

Suppose the right-hand side of (14.5) satisfies the following condition:

$$\varepsilon^2||h||_2 + \varepsilon||h||_1 + ||h|| \leq C, \tag{14.7}$$

where C does not depend on ε. Then the following uniform estimate of $z(\varepsilon)$ holds:

$$\varepsilon^{5/2}||z||_4 + \varepsilon^{3/2}||z||_3 + \varepsilon^{1/2}||z||_2 + ||z||_{3/2} \leq C. \tag{14.8}$$

Proof   We shall look for the solution z of the linear elliptic equation (14.5) in this way (see Vishik & Lusternik [1], [2]):

$$z = z_0 + v + r, \tag{14.9}$$

where z_0 is the solution of the limit problem (for $\varepsilon = 0$)

$$\Delta z_0 = h, \quad z_0|_{\partial\Omega} = 0,$$

$v = \varepsilon c(x)e^{-\rho/\varepsilon}$ is a boundary layer function of first order, which is added to z_0 to satisfy all boundary conditions (14.6) (ρ is the distance to $\partial\Omega$ measured along the normal to $\partial\Omega$, $\rho < \eta$), r is the remainder term. From *a priori* estimates for solutions of linear elliptic equations it follows that the condition (14.7) implies the following estimate:

$$\varepsilon^2||z_0||_4 + \varepsilon||z_0||_3 + ||z_0||_2 \leq C. \tag{14.10}$$

The coefficient $c(x)$ in the expression for v is determined in order to satisfy the second boundary condition (14.6):

$$c(x) = \frac{\partial z_0}{\partial \rho}\Big|_{\rho=0} .$$

It should be noted that (14.10) implies

$$\varepsilon^2 ||c||_{5/2} + \varepsilon ||c||_{3/2} + ||c||_{1/2} \le C_1. \tag{14.11}$$

To satisfy the first boundary condition (14.6) we add to $z_0 + v$ the function $-\varepsilon c(x)$. Finally we multiply the function v_0 (and also the term $-\varepsilon c(x)$), defined in the neighbourhood of boundary $\Omega_\eta = \{\rho < \eta\}$, by an infinitely differentiably smoothing factor $\chi(\rho)$ such as (12.12) to define these functions over the whole domain Ω. It follows from (14.11) that the boundary layer term satisfies the estimate

$$\varepsilon^{1/2} ||v||_2 + \varepsilon^{-1/2} ||v||_1 \le C. \tag{14.12}$$

Applying the interpolation inequality (Agranovich & Vishik [1])

$$\varepsilon^{1-\delta} ||u||_{1-\delta} \le C_\delta (\varepsilon ||u||_1 + ||u||), \quad 0 \le \delta \le 1 \tag{14.13}$$

we get

$$||v||_{3/2} \le C. \tag{14.14}$$

Observe that the remainder $r = z - (z_0 + v - \varepsilon c)$ satisfies homogeneous boundary conditions (14.6). In a similar manner to the proof of Theorem 12.1 in §13, we obtain the following energy estimate, using (14.10), (14.11) and (14.12):

$$\varepsilon^2 ||\Delta r||^2 + ||\nabla r||^2 = |(L_\varepsilon r, r)| \le \frac{1}{2}(\varepsilon^2 ||\Delta r||^2 + ||\nabla r||^2) + C\varepsilon,$$

which implies

$$\varepsilon^{1/2} ||r||_2 + \varepsilon^{-1/2} ||r||_1 \le C, \tag{14.15}$$

$$||r||_{3/2} \le C.$$

Thereby it follows from (14.9), (14.10), (14.11), (14.12) and (14.15) that the solution $z(\varepsilon)$ of the problem (14.5) - (14.6) satisfies the following estimate:

$$\varepsilon^{1/2} ||z||_2 + ||z||_{3/2} \le C. \tag{14.16}$$

Rewriting now equation (14.5) in the form

$$\Delta^2 z = \frac{1}{\varepsilon^2}(\Delta z - h)$$

and taking into account (14.7), (14.6) and the interpolation inequality (14.13), we obtain the estimate (14.8).

The proof of Proposition 13.3 is based on the result of this lemma.

Proof of Proposition 13.3   We differentiate the equation (11.1) with respect to x_i. Taking into consideration the boundary conditions (11.2), we get the following boundary value problem for $p_1(t) = \partial u/\partial x_i$:

$$\partial_t p_1 = -\varepsilon^2 \Delta^2 p_1 + \Delta p_1 - f'(u)p_1 - g'_{x_i},$$
$$p_1|_{\partial\Omega} = \frac{\partial p_1}{\partial y}\,|_{\partial\Omega} = 0. \tag{14.17}$$

We remark that equation (14.17) differs from equation (11.1) for u only in the term $f'(u)p_1$. Multiplying (14.17) by $t^2 p_1$ in L_2, integrating with respect to t, using the Gronwall inequality and taking into account (13.2), we obtain

$$t^2\|p_1(t)\|^2 + \int_0^t (\varepsilon^2\tau^2\|p_1(\tau)\|_2^2 + \tau^2\|p_1(\tau)\|_1^2)\,d\tau \leq C. \tag{14.18}$$

Multiplying (14.17) by $t^4\partial_t p_1$ in L_2 and using the estimate (13.2), we have

$$t^4\|\partial_t p_1\|^2 + \partial_t(t^4\varepsilon^2\|\Delta p_1\|^2 + t^4\|\nabla p_1\|^2)$$
$$\leq Ct^3(\varepsilon^2\|p_1\|_2^2 + \|p_1\|_1^2) + Ct^2\|p_1\|_1^2 + Ct^4.$$

Integrating this inequality with respect to t, and taking (14.18) into account, we get

$$\int_0^t \tau^4\|\partial_t p_1\|^2\,d\tau + t^4\varepsilon^2\|p_1(t)\|_2^2 + t^4\|p_1\|_1^2 \leq C, \tag{14.19}$$

where $t \in [0,T]$ and C depends only on $\|u(0)\|$ and T. Then we differentiate (14.17) with respect to t:

$$\partial_{tt}^2 = -\varepsilon^2\Delta^2(p_1)_t + \Delta(p_1)_t - f'(u)(p_1)_t - f''(u)u_t p_1,$$
$$(p_1)_t|_{\partial\Omega} = \frac{\partial(p_1)_t}{\partial y}\,|_{\partial\Omega} = 0. \tag{14.20}$$

Multiplying (14.20) by $t^5(p_1)_t$ in L_2 and using the Hölder inequality and the estimates (13.2) and (13.3), we obtain

$$\partial_t(t^5\|(p_1)_t\|^2) + t^5\varepsilon^2\|\Delta(p_1)_t\|^2 + t^5\|\nabla(p_1)_t\|^2 \le C(t^2\|p_1\|_1^2 + t^4\|(p_1)_t\|^2).$$

Integrating this inequality with respect to t and using (14.18) and (14.19), we have

$$t^5\|(p_1)_t\|^2 + \int_0^t (\tau^5\varepsilon^2\|(p_1)_t\|_2^2 + \tau^5\|(p_1)_t\|_1^2)\,d\tau \le C. \qquad (14.21)$$

Finally, we multiply equation (14.20) by $t^7(p_1)_{tt}$ in L_2 and then integrate with respect to t. The estimates (13.2), (13.3) and (14.21) imply

$$\int_0^t \tau^7\|(p_1)_{tt}\|^2\,d\tau + t^7\varepsilon^2\|(p_1)_t\|_2^2 + t^7\|(p_1)_t\|_1^2 \le C. \qquad (14.22)$$

Observe that constants C in (14.21) and (14.22) depend only on $\|u(0)\|$ and T. This inequality implies (13.5) for $\ell = 1$.

Hence, the estimates (14.18), (14.19) and (14.22) for $p_1(x, y, t)$, which are similar to the estimates (13.1), (13.2) and (13.3) for $u(x, y, t)$ (for fixed $t > 0$), are established.

Put $t = 1$ in equation (14.17) and rewrite it in the form (14.5) with the right-hand side $h_1(\varepsilon) \equiv \partial_t p_1(1) + f'(u(1))p_1(1) + g'_{x_i}$. We shall prove below that the function h_1 satisfies the condition (14.7) of Lemma 14.1. Indeed, it follows from (12.2), (11.11) and (14.19) that

$$\|f'(u(1))p_1(1)\|^2 \le C(1 + \|u^2 p_1\|^2) \le C(1 + \|u\|_{L_6}^4\|p_1\|_{L_6}^2) \le C \quad (14.23)$$

(we use also the Hölder inequality and the embedding theorems (13.13)). Further, using the estimates (13.2), (14.19) and the embedding $W_2^2(\Omega) \hookrightarrow W_6^1(\Omega)$ we obtain

$$\begin{aligned}
\|f'(u(1))p_1(1)\|_1^2 &= \sum_{i=1}^3 \|f''(u)p_1 D_i u + f'(u)D_i p_1\|^2 + \|f'(u)p_1\|^2 \\
&\le \|f''(u)\|_{L_6}^2\|p_1\|_{L_6}^2\|u\|_{W_6^1}^2 + \|f'(u)\|_{L_3}^2\|p_1\|_{W_6^1}^2 + C \\
&\le C_1(\|u\|_2^2 + \|p_1\|_2^2 + 1) \le C_2\varepsilon^{-2}.
\end{aligned}$$
$$(14.24)$$

Observe that rewriting equation (14.17) (for $t = 1$) as

$$\varepsilon^2\Delta^2 p_1 = \Delta p_1 + \partial_t p_1 - f'(u)p_1 - g'_{x_i}$$

we obtain from (14.19), (14.21) and (14.23) that

$$\|\Delta^2 p_1(1)\| \leq C\varepsilon^{-3},$$

and then the interpolation inequality (13.27) implies the following estimate:

$$\varepsilon^3 \|p_1\|_4 + \varepsilon^2 \|p_1\|_3 + \varepsilon \|p_1\|_2 + \|p_1\|_1 \leq C. \tag{14.25}$$

In exactly the same way, using (13.2) and (13.3), we can prove the analogous estimate for $u = u(1)$:

$$\varepsilon^3 \|u\|_4 + \varepsilon^2 \|u\|_3 + \varepsilon \|u\|_2 + \|u\|_1 \leq C. \tag{14.26}$$

We can estimate now the expression $\|f'(u)p_1\|_2$. Using the Hölder inequality, the estimates (11.11), (14.25), (14.26) and Sobolev embedding theorems, we have

$$\|f'(u)p_1(1)\|_2^2 = \sum_{i,j=1}^{3} \|f'''(u)p_1 D_i u D_j u + 2f''(u)D_i u D_j p_1$$

$$+ f''(u)p_1 D_{ij}^2 u + f'(u)D_{ij}^2 p_1\|^2 + \|f'(u)p_1\|_1^2$$

$$\leq C(\|u\|_{W_6^1}^4 \|p_1\|_{L_6}^2 + \|f''(u)\|_{L_6}^2 \|u\|_{W_6^2}^2 \|p_1\|_{L_6}^2$$

$$+ \|f''(u)\|_{L_6}^2 \|u\|_{W_6^1}^2 \|p_1\|_{W_6^1}^2 + \|f'(u)\|_{L_3}^2 \|p_1\|_{W_6^2}^2 + 1)$$

$$\leq C_1(\|u\|_2^4 \|p_1\|_1^2 + \|u\|_1^2 \|u\|_3^2 \|p_1\|_1^2$$

$$+ \|u\|_1^2 \|u\|_2^2 \|p_1\|_2^2 + \|u\|_1^2 \|p_1\|_3^2 + 1)$$

$$\leq C_2 \varepsilon^{-4}. \tag{14.27}$$

All the constants in these estimates are independent of ε. Observe that the inequality (14.22) implies that the function $\partial_t p_1(1)$ satisfies (14.7).

Hence we have established that the function $h_1(\varepsilon)$ satisfies the condition (14.7) and therefore it follows from (14.8) that the inequality (13.4) is valid for $\ell = 1$.

Remark 14.1.

In a similar manner, taking into account the estimates (13.2), (14.19) and (14.21), we obtain the inequalities for the function $t^4 f'(u(t))p_1(t)$ which are similar to the estimates (14.23), (14.24) and (14.27). It follows from these inequalities that assertion (13.4) is valid also for $t^4 p_1(t)$, $t \in [0, T]$.

The proof of (13.4) for $\ell = 2, 3, \ldots$ is entirely analogous. The equation (14.17) is differentiated with respect to x_i step by step and the estimates, similar to (14.18), (14.19) and (14.22), are established:

$$\int_0^t \tau^{m_1} \|\partial_t p_\ell\|^2 \, d\tau + t^{m_1} \varepsilon^2 \|p_\ell(t)\|_2^2 + t^{m_1} \|p_\ell(t)\|_1^2 \leq C, \quad (14.28)$$

$$\int_0^t \tau^{m_2} \|(p_\ell)_{tt}\|^2 \, d\tau + t^{m_2} \varepsilon^2 \|(p_\ell)_t\|_2^2 + t^{m_2} \|(p_\ell)_t\|_1^2 \leq C. \quad (14.29)$$

We use here the embedding theorems (13.13) and boundedness of derivatives $f^{(\ell)}(u)$ for $\ell \geq 3$.

For instance, the function $p_2(t)$ satisfies the following equation:

$$\partial_t p_2 = -\varepsilon^2 \Delta^2 p_2 + \Delta p_2 - f'(u)p_2 - f''(u)p_1^2 - g''_{x_i}, \quad (14.30)$$

which differs from equation (14.17) only in the term $f''(u)p_1^2$. We can estimate this term thus:

$$t^4 \|f''(u)p_1^2\| \leq t^4 \|f''(u)\|_{L_3} \|p_1^2\|_{L_{3/2}} \leq C t \|u\|_1^2 t^3 \|p_1(t)\|_{L_3}^2 \leq C$$

(we use estimate (13.2) and inequalities (14.18) and (14.19), which imply that $t^{3/2} \|p_1(t)\|_{1/2} \leq C$). Then we differentiate equation (14.30) with respect to t. All terms in the expression obtained except $f'''(u)u_t p_1^2$ are similar to the terms in (14.20). For this new term we deduce

$$t^6 \|f'''(u)u_t p_1^2\| \leq C t^6 \|u_t p_1^2\| \leq C t^3 \|u_t\|_1^2 t^3 \|p_1(t)\|_{L_3}^2 \leq C$$

(in accordance with (13.3)). Hence, as before, we obtain the estimates (14.28), (14.29) for $\ell = 2$.

For any ℓ these estimates are established in a similar manner.

Remark 14.2.

In Theorem 12.2 a stronger result than estimate (12.17) is valid (if all the functions are sufficiently smooth):

$$\|u\|_{H_s(\Omega_\gamma)} \leq M,$$

where $s \geq 3$ and $M = M(\gamma, s)$ doesn't depend on $u \in \mathcal{A}^\varepsilon$ or on $\varepsilon \in (0, \varepsilon_0]$ (Vishik & M. Skvortsov [2], M. Skvortsov & Vishik [1]). Similarly in Theorem 12.3 we can replace the metrics $H_{3-\delta}(\Omega_\gamma)$ by $H_{s-\delta}(\Omega_\gamma)$, $s \geq 3$.

Remark 14.3.

Let $\{u(t)\}$ be a set of trajectories of the problem (11.1) - (11.2) with bounded initial data:

$$\|u|_{t=0}\| \leq R.$$

Then the asymptotic expansion (11.9) holds (for fixed $T > 0$):

$$u(T) = u_0 + v_0 - \varepsilon c_0 + \varepsilon(u_1 + v_1 - \varepsilon c_1) + r.$$

The remainder r satisfies estimate (11.10), in which M depends only on R and T, and the functions u_0, v_0, c_0, u_1, v_1, c_1 satisfy the same estimates as above with constants $C = C(R, T)$.

§15 Problem with boundary conditions $u|_{\partial\Omega} = 0$, $\Delta u|_{\partial\Omega} = 0$

Let the function u be the solution of equation (11.1) and satisfy the boundary conditions

$$u|_{\partial\Omega} = 0, \quad \Delta u|_{\partial\Omega} = 0. \tag{15.1}$$

In Chapter VII the global asymptotics of the solution $u(t)$ of (11.1), (15.1) will be established for all t.

Proposition 15.1.

If g belongs to $H_1(\Omega)$, $f(0) = 0$ and $f(u)$ satisfies the conditions (12.1) - (12.4) ($\ell \geq 3$ in (12.3)), then the following estimates of the solution $u(t)$ of the problem (12.1), (15.1) hold:

$$\varepsilon^2 t^2 \|u(t)\|_3^2 + t^2 \|u(t)\|_2^2 + \int_0^t \tau^2 \|u_t(\tau)\|_1^2 \, d\tau \leq C, \tag{15.2}$$

$$\varepsilon^2 t^6 \|u_t(t)\|_3^2 + t^6 \|u_t(t)\|_2^2 + \int_0^t \tau^6 \|u_{tt}(\tau)\|_1^2 \, d\tau \leq C, \tag{15.3}$$

where $t \in [0, T]$ and the constants C depend only on $\|u(0)\|$ and T. Besides,

$$\varepsilon^2 \|u(t)\|_3^2 + \|u(t)\|_2^2 \leq C_1(\delta), \quad 0 < \delta \leq t < +\infty. \tag{15.4}$$

Proof   Obviously $u(t)$ satisfies estimates (13.1) - (13.3). Let us multiply equation (11.1) by $t\Delta u$ in L_2,

$$\frac{1}{2}\partial_t(t||\nabla u||^2) + \varepsilon^2 t||\nabla(\Delta u)||^2 + t||\Delta u||^2 + t(f'(u)\nabla u, \nabla u)$$
$$= \frac{1}{2}||\nabla u||^2 + t(g, \Delta u),$$

and then integrate with respect to t. Using (13.1) and (13.2) we get

$$\int_0^t (\varepsilon^2 \tau ||u(\tau)||_3^2 + \tau ||u(\tau)||_2^2)\,d\tau \le C. \tag{15.5}$$

Now multiply equation (11.1) by $t^2 \Delta u_t$:

$$t^2||\nabla u_t||^2 + \frac{1}{2}(\varepsilon^2 \partial_t(t^2||\nabla(\Delta u)||^2) + \partial_t(t^2||\Delta u||^2)) \tag{15.6}$$
$$\le \varepsilon^2 t||\nabla(\Delta u)||^2 + t||\Delta u||^2 + t^2|(f'(u)\nabla u, \nabla u_t)| + t^2|(\nabla g, \nabla u_t)|.$$

It follows from estimate (13.2) and condition (12.2) that

$$t^2|(f'(u)\nabla u, \nabla u_t)| \le Ct^2||f'(u)||_{L_3}||\nabla u||_{L_6}||\nabla u_t|| \tag{15.7}$$
$$\le C_1||u||_1^2 + \frac{1}{3}t^2||\nabla u_t||^2.$$

Therefore, integrating (15.6) with respect to t and using (13.1) and (15.5), we obtain inequality (15.2).

Multiplying equation (14.3) by $t^3 \Delta u_t$ in L_2 and integrating with respect to t, we get

$$\int_0^t \tau^3(\varepsilon^2||u_t||_3^2 + ||u_t||_2^2)\,d\tau \le C. \tag{15.8}$$

Finally, we multiply equation (14.3) by $t^6 \Delta u_{tt}$

$$t^6||\nabla u_{tt}||^2 + \frac{1}{2}\partial_t[t^6 \varepsilon^2||\nabla(\Delta u_t)||^2 + t^6||\Delta u_t||^2] \tag{15.9}$$
$$= 3(\varepsilon^2 t^5||\nabla(\Delta u_t)||^2 + t^5||\Delta u_t||^2) + t^6(f'(u)u_t, \Delta u_{tt}).$$

Observe that inequality (15.2) implies that

$$t|f''(u)| \le Ct(|u|+1) \le Ct(||u||_2 + 1) \le C_1$$

since $H_2 \subseteq C$ for $n = 3$. The last term in the right-hand side of (15.9) is estimated similarly to (15.7):

$$
\begin{aligned}
t^6|(f'(u)u_t, \Delta u_{tt})| &= t^6|(f''(u)u_t\nabla u + f'(u)\nabla u_t, \nabla u_{tt})| \\
&\leq (t\|f''(u)\|_C)(t\|u_t\|_{L_3})(t\|u\|_2)(t^3\|\nabla u_{tt}\|) \\
&\quad + (t\|f'(u)\|_{L_3})(t^2\|u_t\|_2)(t^3\|\nabla u_{tt}\|) \\
&\leq C_1 t^2\|u_t\|_1^2 + C_2 t^4\|u_t\|_2^2 + \frac{1}{2}t^6\|\nabla u_{tt}\|^2.
\end{aligned}
$$

Integrating (15.9) with respect to t and using (15.8) we get estimate (15.3). To prove (15.4) we observe that from (14.1) and (12.1) it follows that

$$\|u(t)\|^2 \leq \|u(0)\|^2 \exp(-\gamma^2 t) + \frac{C}{\gamma^2}(1 - \exp(-\gamma^2 t)).$$

Hence $\|u(t)\|$ is bounded for all $t \geq 0$ if $\|u(0)\|$ is bounded. As has been shown above the constant C in (15.2) depends only on $\|u(0)\|$ and T. So applying estimate (15.2) to any interval $(t_0, t_0 + T)$ and taking into account that $\|u(t_0)\| \leq M \ \forall t_0 > 0$ we deduce (15.4).

We can obtain now the asymptotic expansion of elements $u(T)$ of sets of solutions $\{u(t)\}$ which is similar to asymptotics (11.9) of solutions of the problem (11.1) - (11.2). Henceforth we shall need only the first terms of expansion (11.9).

Proposition 15.2.
Let g belong to $H_2(\Omega)$ and suppose all conditions of Proposition 15.1 are satisfied. Let $\{u(t)\}$ be a set of trajectories of the problem (11.1), (15.1) with bounded initial data:

$$\|u|_{t=0}\| \leq R.$$

Then for any $T > 0$ the following expansion holds:

$$u(T) = u_0 + v - \varepsilon^2 c + r, \tag{15.10}$$

where v is the second-order boundary layer function (i.e. function of type $\varepsilon^2 C \exp(-y/\varepsilon)$) and the following estimates are valid:

$$\varepsilon\|u_0\|_5 + \|u_0\|_4 \leq C_1, \tag{15.11}$$

$$\varepsilon\|c\|_{5/2} + \|c\|_{3/2} \leq C_2, \tag{15.12}$$

$$\varepsilon^{1/2}\|v\|_{5/2} + \|v\|_2 + \varepsilon^{-1}\|v\|_1 + \varepsilon^{-2}\|v\| \leq C_3\varepsilon^{1/2}, \tag{15.13}$$

$$\|r\|_2 + \varepsilon^{-1}\|r\|_1 \leq C_4\varepsilon; \tag{15.14}$$

here all the constants depend on T and R only.

Proof   As above the function u_0 is the solution of the problem (12.6), (12.7):

$$\Delta u_0 = h \equiv \partial_t u(T) + f(u(T)) + g, \quad u_0|_{\partial\Omega} = 0.$$

v is the second order boundary layer function (which is necessary to satisfy the second of the boundary conditions (15.1)) and in the neighbourhood of the boundary $\Gamma_0 = \partial\Omega \cap \{y = 0\}$ has the form

$$\tilde{v} = \varepsilon^2 \tilde{c}(x) e^{-y/\varepsilon}, \quad \tilde{c}(x) = -\frac{\partial^2 u_0}{\partial y^2}\Big|_{y=0} \tag{15.15}$$

(compare with (12.10), (12.11)). The functions $v(x,y)$ and $c(x,y)$ have similar form near the boundary $\Gamma_1 = \partial\Omega \cap \{y = 1\}$ and are determined in the interior of Ω with the aid of the smoothing function $\chi(y)$ (see (12.12)).

From the definitions of functions u_0, v and c it follows that the remainder r satisfies homogeneous boundary conditions (15.1).

Since

$$D^2 f(u) = f''(u)(Du)^2 + f'(u)D^2 u,$$
$$D^3 f(u) = f'''(u)(Du)^3 + 3f''(u)DuD^2 u + f'(u)D^3 u$$

and $u(T)$ belongs to $H_2(\Omega)$ and $C(\Omega)$ if $T > 0$, it follows from (15.2) and (15.3) that the right-hand side h of equation (12.6) is bounded in $H_2(\Omega)$ uniformly with respect to ε and εh is uniformly bounded in $H_3(\Omega)$. Hence estimate (15.11) is valid for the solution u_0 of the problem (12.6) - (12.7) and estimate (15.12) is valid for the function $c(x,y)$, and constants C_1 and C_2 in these estimates depend only on T and $\|u(0)\|$.

Consider the expression

$$L_\varepsilon r \equiv -\varepsilon^2 \Delta^2 r + \Delta r = \varepsilon^2 \Delta^2 u_0 - \varepsilon^4 \Delta^2 c - L_\varepsilon v.$$

Multiplying it by r in L_2 and using inequalities similar to estimates (13.17), (13.18), (13.20) and (13.23) we get

$$\varepsilon^2 \|\Delta r\|^2 + \|r\|_1^2 \leq C\varepsilon^4.$$

From the form (15.15) of the boundary layer function v and inequality (15.12) follows estimate (15.13).

VII

Asymptotics of solutions of the singular perturbed evolutionary equations

In this chapter, written by M. Yu. Skvortsov and M. I. Vishik, their new results are presented.

§16 Asymptotics of trajectories of the first boundary value problem

We shall be concerned with the singular perturbed reaction-diffusion parabolic system in the bounded domain Ω in $\mathbf{R}^n$:

$$\partial_t u = -\varepsilon^2 \Delta^2 u + a\Delta u - f(u) - g. \qquad (16.1)$$

Here $u = (u^1, ..., u^m)$, $f = (f^1, ..., f^m)$, $g = (g^1, ..., g^m)$, a is a matrix of order m with constant coefficients which has positive symmetrical part: $a + a^* > 0$. For simplicity we consider the scalar equation ($m = 1$); it is not difficult to extend our results to any $m > 1$. We assume that $a = 1$ for brevity. On the boundary $\partial\Omega$ the boundary conditions

$$u\,|_{\partial\Omega} = 0, \quad \frac{\partial u}{\partial \nu}\,|_{\partial\Omega} = 0 \qquad (16.2)$$

are given. Let g belong to $H_2(\Omega)$, $f(0) = 0$ and the restrictions (12.1) - (12.4) on the function $f(u)$ be satisfied ($\ell \geq 3$ in (12.3)). As in §12 we assume that $n = 3$ and Ω is the cylinder domain $\Omega = T_x^2 \times \{0 < y < 1\}$. Below we shall obtain the principal term of the asymptotics of the solution $u_\varepsilon(t) = u(t)$ of the system (16.1) on any finite interval $[0, T]$.

For $\varepsilon = 0$ the problem (16.1) - (16.2) reduces to the limit problem

$$\partial_t w = \Delta w - f(w) - g, \quad w|_{\partial\Omega} = 0. \qquad (16.3)$$

Denote by $U(R) = \{u(t)\}$ the set of the trajectories of the dynamical system (16.1) - (16.2) defined for $-\gamma \leq t \leq T$ and having bounded initial data:

$$\|u(-\gamma)\| \leq R, \quad \gamma > 0. \tag{16.4}$$

Then, according to Remark 14.3, the asymptotic expansion (11.9) holds for the functions $u(0)$:

$$u(0) = u_0 + v_0 - \varepsilon c_0 + \varepsilon(u_1 + v_1 - \varepsilon c_1) + r \equiv u_0 + v^0, \tag{16.5}$$

where u_0 is the solution of the problem (12.6) - (12.7) and

$$v^0 \equiv v_0 - \varepsilon c_0 + \varepsilon u_1 + \varepsilon v_1 - \varepsilon^2 c_1 + r.$$

From corresponding estimates of each term of (16.5) it follows that

$$\|u_0\|_3 \leq C_1 \tag{16.6}$$

and

$$\|v^0\| \leq C_2\varepsilon, \quad \|v^0\|_3 \leq C_3\varepsilon^{-3/2}, \tag{16.7}$$

where the constants C_1, C_2 and C_3 depend on γ and R only.

It should be noted that since equation (16.1) is autonomous, all the following assertions are valid on the interval $[t_0 - \gamma, t_0 + T]$ for any $t_0 \in \mathbf{R}_+$.

Remark 16.1.

We can set $U(R) = \mathcal{A}^\varepsilon$, where $\mathcal{A}^\varepsilon$ is the attractor, corresponding to the problem (16.1) - (16.2). Hence all the following is valid for the trajectories $u(t)$ belonging to $\mathcal{A}^\varepsilon$.

Theorem 16.1.

For the trajectories $u(t) \in U(R)$ the following asymptotic expansion, uniform for $\varepsilon \in (0, \varepsilon_0]$, holds:

$$u(t) = w(t) + v(t) + r(t); \tag{16.8}$$

here $t \in [0, T]$, $w(t)$ is the solution of the limit equation (16.3) with initial data $w(0) = u_0$ (the function u_0 has been defined in (16.5)), $v(t)$ is the first order boundary layer function and the remainder $r(t)$ satisfies the following estimates:

$$\|r(t)\|^2 + \int_0^t (\varepsilon^2\|r\|_2^2 + \|r\|_1^2)\,d\tau \leq C_1\varepsilon^2, \tag{16.9}$$

$$\int_0^t \tau^m ||\partial_t r||^2 \, d\tau + \varepsilon^2 t^m ||r(t)||_2^2 + t^m ||r(t)||_1^2 \le C_2 \varepsilon^2, \qquad (16.10)$$

where $m > 0$ and the constants $C_1(T, R)$ and $C_2(T, R)$ do not depend on ε or t.

Proof   First we shall formulate some estimates for the solution $w(t)$ of problem (16.3).

Proposition 16.1.
     Let the function g belong to $H_2(\Omega)$ and $w(t)$ be the solution of the problem (16.3) with the initial data $w(0) \in H_3(\Omega)$. Then the following estimate holds for $t \in [0, T]$:

$$||w(t)||_3^2 + ||w_t(t)||_1^2 + \int_0^t (||w||_4^2 + ||w_{tt}||^2 + ||w_t||_2^2) \, d\tau \le C. \qquad (16.11)$$

If, furthermore, the derivatives $\partial^j g / \partial x_i^j$ $(i = 1, 2; \, j = 1, 2, 3...)$ belong to $H_2(\Omega)$, then

$$t^{m_j} ||(p_j(t))_t||_1^2 + t^{m_j} ||p_j(t)||_4^2 + \int_0^t \tau^{m_j} ||(p_j)_{tt}||^2 \, d\tau \le C, \qquad (16.12)$$

where $p_j = \partial^j w / \partial x_i^j$, $m_j > 0$ and all the constants $C = C(T)$ are independent of t.

     These estimates will be proved below.
     The asymptotic expansion (16.8) is constructed similarly in many respects to the stationary asymptotics of an element u of trajectory $u(t)$ which has been established in §§12 - 15. In the neighbourhood of $\Gamma_0 = \partial\Omega \cap \{y = 0\}$ the function v has the form

$$v = v_0 = \varepsilon c_0(x, y) e^{-y/\varepsilon}, \quad \text{where } c_0(x, t) = \frac{\partial w}{\partial y} \Big|_{y=0}$$

(compare with (12.9), (12.10)). Similarly the function v is determined in the neighbourhood of $\Gamma_1 = \partial\Omega \cap \{y = 1\}$ (replacing c_0 by c_1), and then v is extended to the whole of the domain Ω with the aid of the smoothing function $\chi(y)$, which has been defined in (12.12). Let $c(x, y, t)$ denote the function which is equal to c_0 in the neighbourhood of Γ_0, equal to c_1 in the neighbourhood of Γ_1, and extended to Ω with the aid of the smoothing function. Then the function $w + v - \varepsilon c$ satisfies the boundary conditions (16.2).

Substituting the asymptotic expansion

$$u = w + v - \varepsilon c + r$$

into equation (16.1), we obtain the following problem for determination of the remainder $r(t)$:

$$\partial_t r = -\varepsilon^2 \Delta^2 r + \Delta r - d(u,w)r + R(u,w),$$

$$r\,|_{\partial\Omega} = \frac{\partial r}{\partial y}\,|_{\partial\Omega} = 0, \qquad (16.13)$$

$$r|_{t=0} = v^0 - (v - \varepsilon c)|_{t=0}, \qquad (16.14)$$

where

$$d(u,w) \equiv \int_0^1 f'(w + \theta(u - w))\, d\theta,$$

$$R(u,w) \equiv -\partial_t(v - \varepsilon c) - \varepsilon^2 \Delta^2 w + (-\varepsilon^2 \Delta^2 v + \Delta v)$$
$$+ \varepsilon^3 \Delta^2 c - \varepsilon \Delta c - d(u,w)(v - \varepsilon c).$$

Observe that it follows from (16.11) that the function $q = \partial w / \partial y$ satisfies the following estimate:

$$\|q\|_2^2 + \|q_t\|^2 + \int_0^t (\|q\|_3^2 + \|q_t\|_1^2)\, d\tau \le C;$$

this implies that

$$c(x_1, x_2, y, t) \in W_2^{\ell}(Q) \qquad (16.15)$$

where $Q = \Omega \times (0, T)$, $\ell = (5/2, 5/2, \infty, 1)$.

It follows from the embedding theorem for non-isotropic spaces (13.7) that

$$\|c|_{t=0}\|_{H_{5/4}(\Omega)} \le C. \qquad (16.15')$$

Hence (16.14) and (16.5) imply

$$\|r|_{t=0}\| = \|v^0 - v(0) + \varepsilon c(0)\| \le C\varepsilon. \qquad (16.16)$$

Multiply equation (16.13) by r in L_2:

$$\frac{1}{2}\partial_t \|r\|^2 + \varepsilon^2 \|\Delta r\|^2 + \|\nabla r\|^2 = -(dr, r) + (R, r). \qquad (16.17)$$

Since $f'(u) \geq -C$,

$$-(dr,r) = -\int_0^1 (f'(w + \theta(u - w)r, r)\, d\theta \leq C\|r\|^2.$$

We estimate now each term in the expression (R, r). From the form of the boundary layer function v and (16.15) it follows that

$$|(-\partial_t(-v + \varepsilon c), r)| \leq C_1\varepsilon|(\partial_t c, r)| \leq C_2\varepsilon^2 + \|r\|^2$$

and

$$|(-\varepsilon^2\Delta^2 v + \Delta v, r)| \leq \left| \left(-\varepsilon^2\frac{\partial^2 v}{\partial y^2} + v, \Delta r\right) \right| + \varepsilon^2|(\Delta_x v, \Delta r)|$$
$$\leq C_3\varepsilon^4 + \mu\varepsilon^2\|\Delta r\|^2$$

(see, also, (13.18) - (13.23)). Then

$$\varepsilon^2|(\Delta^2 w, r)| \leq \varepsilon^4\|w\|_4^2 + \|r\|^2,$$
$$\varepsilon^3|(\Delta^2 c, r)| = \varepsilon^3|(\Delta c, \Delta r)| \leq C_4\varepsilon^4 + \mu\varepsilon^2\|\Delta r\|^2,$$
$$\varepsilon|(\Delta c, r)| \leq C_5\varepsilon^2 + \|r\|^2$$

and, finally,

$$|(d(v - \varepsilon c), r)| \leq C\|d\|_{L_3}\|v - \varepsilon c\|_{L_6}\|r\| \leq C_6\varepsilon^2(1 + \|u\|_1^2) + \|r\|^2 \quad (16.18)$$

(we use here condition (12.2) on the growth of $f'(u)$ and estimate (16.15') for the function c). Substituting these estimates into (16.17), choosing μ sufficiently small, integrating with respect to t and using (16.16) and the energy estimate (13.1), we deduce (16.9).

To prove inequality (16.10) we multiply equation (16.13) by $t^m r_t$ in L_2:

$$t^m\|r_t\|^2 + \frac{1}{2}\varepsilon^2\partial_t(t^m\|\Delta r\|^2) + \frac{1}{2}\partial_t(t^m\|\nabla r\|^2)$$
$$= \frac{m}{2}t^{m-1}(\varepsilon^2\|\Delta r\|^2 + \|\nabla r\|^2) - t^m(dr, r_t) + t^m(R, r_t). \quad (16.19)$$

The term $t^m(dr, r_t)$ is estimated similarly to (16.18). The terms in the expression (R, r_t) are estimated also as the terms of (R, r) above. We note only that

$$-\varepsilon^2\Delta^2 v + \Delta v = \Delta(-\varepsilon^2\partial_{yy}^2 v + v - \varepsilon^2\Delta_x v)$$
$$= \varepsilon^3\Delta\left[\left(\chi'' - 2\frac{\chi'}{\varepsilon}\right)ce^{-y/\varepsilon}\right] - \varepsilon^2\Delta_x^2 v - \varepsilon^2\partial_{yy}^2\Delta_x v \quad (16.20)$$

(see also (13.23)). Then, since inequality (16.12) for $j = 2$ implies that

$$q(t) = t^m \frac{\partial w}{\partial y} \in W_2^{\ell}(\Omega),$$

where $\ell = (5, 5, 3)$, so, using the embedding theorem (13.7), in which $\psi = 5/6$, we get

$$t^m \frac{\partial w}{\partial y} \Big|_{y=0} \in W_2^{25/6}(T_x^2)$$

and hence

$$t^m \|\Delta^2 c\| \leq C t^m \|\partial_y w(x, 0, t)\|_{H_4(T_x^2)} \leq C_1.$$

This inequality and (16.20) imply that

$$t^m \| - \varepsilon^2 \Delta^2 v + \Delta v \| = \mathcal{O}(\varepsilon \|\Delta_x c \cdot e^{-y/\varepsilon}\|) = \mathcal{O}(\varepsilon^{3/2})$$

and

$$t^m |(R, r_t)| \leq C_4 \varepsilon^2 + C_5 t^m \|r_t\|^2.$$

Integrating (16.19) with respect to t we get estimate (16.10).

Observe that (16.15) implies that the term $-\varepsilon c$ satisfies estimates (16.9) and (16.10), so it may be included in the remainder. Thus the asymptotic expansion (16.8) is established.

Proof of Proposition 16.1   Multiplying equation (16.3) by w in L_2, then by Δw and by w_t, we deduce the following energy estimate for the solution $w(t)$ of problem (16.3):

$$\|w(t)\|_1^2 + \int_0^t (\|\Delta w\|^2 + \|w_t\|^2) \, d\tau \leq C. \tag{16.21}$$

Put $t = 0$ in equation (16.3):

$$w_t(0) = \Delta w(0) - f(w(0)) - g. \tag{16.22}$$

Since $w(0) = u_0 \in H_3(\Omega)$ and $f(u_0) \in H_1(\Omega)$ and for the function u_0 the compatibility conditions are fulfilled,

$$u_0 \big|_{\partial\Omega} = (\Delta u_0 - g) \big|_{\partial\Omega} = 0,$$

(see (12.6)), so $w_t(0) \in H_1^0(\Omega)$.

Let us differentiate both sides of equation (16.3) with respect to t:

$$w_{tt} = \Delta w_t - f'(w)w_t. \tag{16.23}$$

Multiplying this equation by w_t in L_2, then by Δw_t and by w_{tt} and using estimate (16.21), we deduce

$$||w_t||_1^2 + \int_0^t (||\Delta w_t||^2 + ||w_{tt}||^2)\,d\tau \le C. \tag{16.24}$$

We then rewrite (16.3) as the stationary equation

$$\Delta w = w_t + f(w) + g = h. \tag{16.25}$$

(16.21) and (16.24) imply that the right-hand side h is bounded in $L_2(\Omega)$, hence

$$||w(t)||_2 \le C \quad \forall t \in [0,T].$$

This estimate implies, in its turn, that h is bounded in $H_1(\Omega)$ and so

$$||w(t)||_3 \le C,$$

hence $||f(w)||_2 \le C$.

Then (16.24) implies that

$$h \in L_2((0,T); H_2(\Omega)),$$

hence

$$w \in L_2((0,T); H_4(\Omega)),$$

and so inequality (16.11) is entirely proved.

To prove estimates (16.12) we differentiate (16.23) with respect to t:

$$w_{ttt} = \Delta w_{tt} - f''(w)(w_t)^2 - f'(w)w_{tt}. \tag{16.26}$$

Multiplying this equation by tw_{tt} in L_2 we obtain

$$t||w_{tt}||^2 + \int_0^t \tau||w_{tt}(\tau)||_1^2\,d\tau \le C.$$

Multiplying equation (16.23) by t and rewriting it as a stationary equation similarly to (16.25), we get that the function $tw_t(t)$ is bounded in $H_2(\Omega)$, and then it follows from (16.25) that

$$t||w(t)||_4 \le C.$$

Thus we establish inequality (16.12) for $j = 0$.

The proof of (16.12) for $j = 1, 2, \ldots$ is similar to that of Proposition 13.3. Equation (16.3) is differentiated with respect to x_i step by step. The equations obtained for $p_j = \partial^j w / \partial x_i^j$ $(j = 1, 2, \ldots)$ differ from the previous equations for p_k $(k < j)$ only by terms of smaller order. So estimates (16.12) are established in a similar manner.

Thus, Theorem 16.1 is proved.

Remark 16.2.

The asymptotic expansion (16.8) is also valid for singular perturbed parabolic systems of the type (16.1), $m > 1$. Conditions (12.1) - (12.4) on $f(u)$ are easily reformulated for vector functions. Then $w(t)$ is the solution of the limit system, and the remainder $r = (r^1, \ldots, r^m)$ satisfies the estimate (16.9) (all norms are in spaces $(H_k(\Omega))^m$). Let, furthermore, the matrix a be symmetrical and $f^k = \partial F / \partial u^k$, where F is a certain function. Then the estimate (16.10) is valid.

Let the set $\mathcal{M} = \{z_1, \ldots, z_N\}$ of the equilibrium points of the semigroup $\{S_t(0)\}$ corresponding to the limit problem (16.3) be finite. Denote by $\mathcal{O}_\delta(\mathcal{M})$ a δ-neighbourhood of $\mathcal{M}$ in $H_1(\Omega)$. As above, $U(R) = \{u^\varepsilon(t)\}$ is the set of trajectories of the problem (16.1) - (16.2) such that $\|u^\varepsilon(-\gamma)\| \leq R$, $\gamma > 0$, $\varepsilon \in (0, \varepsilon_0]$.

Proposition 16.2.

Let $u(t) \in U(R)$ and $u(t_1) \notin \mathcal{O}_\delta(\mathcal{M})$. Let t_2 be the moment $(t_2 > t_1)$ when the trajectory $u(t)$ gets into the set $\mathcal{O}_\delta(\mathcal{M})$ for the first time. Then $t_2 - t_1 \leq T$ and $T = T(\delta, R)$ doesn't depend on ε, $u(t)$, t_1, t_2.

Proof   (See also the proof of Theorem 5.2.) Assume the contrary, i.e. there exists $\delta > 0$ such that for any T and for any $\varepsilon_0 > 0$ there exist $\varepsilon \in (0, \varepsilon_0]$ and $u \in u^\varepsilon(t) \in U(R)$ such that

$$u^\varepsilon(t) \equiv S_t(\varepsilon)u \notin \mathcal{O}_\delta(\mathcal{M}) \quad \text{for any } t \in [0, T].$$

We remark that the solution $u(t) \in U(R)$ of equation (16.1) satisfies the following inequalities (see Babin & Vishik [1]):

$$\varepsilon^2 \|\Delta u(t)\|^2 + \|u(t)\|_1^2 \leq C \quad \text{and} \quad \int_0^\infty \|\partial_t u\|^2 \, d\tau \leq C. \tag{16.27}$$

This implies that there exists $t \in [0, T]$ such that

$$\|\partial_t u(t)\| \leq \frac{1}{T}. \tag{16.28}$$

Choosing sequences $T = T_j \to +\infty$ and $\varepsilon_0 = \varepsilon_{0j} \to 0$, and for every T_j choosing $t = t_j$ which satisfies (16.28), we get the sequence $u_j \equiv u^{\varepsilon_j}$ such that

$$u_j(t_j) \notin \mathcal{O}_\delta(\mathcal{M}) \quad \text{and} \quad \|\partial_t u_j(t_j)\| \le \frac{1}{T_j}. \tag{16.29}$$

The sequence $\{u_j(t_j)\}$ is uniformly bounded for ε_j and for j in $H_1(\Omega)$, hence we can choose a sub-sequence $\{u_k(t_k) = S_{t_k}(\varepsilon_k)u_k\}$ which converges to a function u^0 weakly in H_1.

Rewrite equation (16.1) for $t = t_k$, $\varepsilon = \varepsilon_k$ in the form

$$\Delta u_k(t_k) = \partial_t u_k(t_k) + \varepsilon^2 \Delta^2 u_k(t_k) + f(u_k(t_k)) + g \equiv h_k. \tag{16.30}$$

Using the condition on the growth of $f(u)$ and boundedness of $u(t)$ in $H_1(\Omega)$ we obtain that the set $\{f(u_k(t_k))\}$ is uniformly bounded for ε_k and k in $L_2(\Omega)$, and, hence, we can choose a sub-sequence $\{k'\}$ such that

$$f(u_{k'}(t_{k'})) \to f(u^0) \quad \text{weakly in } L_2(\Omega) \tag{16.31}$$

as $k' \to +\infty$ (see Lions [1]).

Moreover, the asymptotic expansion (16.5) (where $u(0)$ is replaced by $u_k(t_k)$), which is uniform for all elements of $U(R)$, implies that

$$\varepsilon_k^2 \Delta^2 u_k(t_k) \to 0 \quad \text{in } H_{-1}(\Omega) \tag{16.32}$$

as $k \to +\infty$ (and $\varepsilon_k \to 0$). Indeed, from estimates (16.6) and (16.7) it follows that if $\eta \in H_1^0(\Omega)$ then

$$(\varepsilon^{5/3}\Delta^2 u, \eta) = -\varepsilon^{5/3}(D^3 u_0, D\eta) - \varepsilon^{5/3}(D^3 v^0, D\eta) \to 0 \quad \text{as } \varepsilon \to 0.$$

Thus, using (16.29), (16.31) and (16.32) we deduce that the right-hand side $h_{k'}$ of equation (16.30) converges to $f(u^0) + g$ in $H_{-1}(\Omega)$ as $k' \to \infty$. Since the operator Δ^{-1} is bounded from H_{-1} to H_1,

$$u_{k'}(t_{k'}) \to u^0 \quad \text{in } H_1(\Omega).$$

So from equation (16.30) we obtain (setting $k' \to +\infty$) that $u^0 \in \mathcal{M}$ which contradicts the assumption $u_{\varepsilon_k}(t_k) \notin \mathcal{O}_\delta(\mathcal{M})$.

This proposition and Theorem 16.1 imply the following.

Theorem 16.2.
Let the trajectory $u^\varepsilon(t)$ belong to the set $U(R)$, the conditions formulated above be fulfilled and $\delta > 0$. Let $u^\varepsilon(t) \in U(R) - \mathcal{O}_\delta(\mathcal{M})$ for $t_1 < t < t_2$. Then
  (1) $t_2 - t_1 \le T$, where $T = T(\delta, R)$ doesn't depend on ε, t_1, t_2 and $u(t) \in U(R)$,
  (2) for $t \in (t_1, t_2)$ the asymptotic expansion (16.8) of the trajectory $u^\varepsilon(t)$ holds, where $w(t)$ is the trajectory of the semigroup $\{S_t(0)\}$, $v(t)$ is the first order boundary layer function and the remainder satisfies the following estimate:

$$\varepsilon^2 \|r(t)\|_2^2 + \|r(t)\|_1^2 + \int_{t_1}^t \|\partial_t r\|^2 \, d\tau \le C\varepsilon^2,$$

where $C = C(R, \delta)$ doesn't depend on ε and $u^\varepsilon(t) \in U(R)$.

§17   Global asymptotics of solutions of singular perturbed equation

We shall establish below the principal term of asymptotics of the solutions $u^\varepsilon(t)$ for all $t \in [0, +\infty)$. We assume here that the boundary conditions are the following:

$$u|_{\partial\Omega} = 0, \quad \Delta u|_{\partial\Omega} = 0. \tag{17.1}$$

First we shall determine asymptotic expansions of the solutions of the problem (16.1), (17.1) on a finite interval of time $[0, T]$. Then the asymptotics will be established for all t.

The following theorem, similar to Theorem 16.1, holds.

Theorem 17.1.
For the trajectories $u(t) \in U(R)$, $t \in [0, T]$, the asymptotic expansion (16.8) holds, where $w(t)$ is the solution of the limit equation (16.3), $v(t)$ is the second order boundary layer function and the remainder $r(t)$ satisfies the following estimates, uniform for ε and t:

$$\|r(t)\|^2 + \int_0^t (\varepsilon^2 \|r\|_2^2 + \|r\|_1^2) \, d\tau \le C\varepsilon^4, \tag{17.2}$$

$$\varepsilon^2 t^m \|r(t)\|_2^2 + t^m \|r(t)\|_1^2 + \int_0^t (\tau^m \|\partial_t r\|^2 + \varepsilon^2 \tau^m \|r\|_3^2 + \tau^m \|r\|_2^2) \, d\tau \le C\varepsilon^4. \tag{17.3}$$

Proof   Proposition 15.2 implies that for the functions $u(0)$ an expansion similar to (16.5) holds:

$$u(0) = u_0 + v - \varepsilon^2 c + r \equiv u_0 + v^0, \qquad (17.4)$$

where u_0 and v^0 satisfy the following estimates:

$$\varepsilon||u_0||_5 + ||u_0||_4 \leq C_1, \qquad (17.5)$$

$$||v^0|| \leq C_2\varepsilon^2, \quad ||v^0||_1 \leq C_3\varepsilon^{3/2}, \quad ||v^0||_2 \leq C_4\varepsilon^{1/2}. \qquad (17.6)$$

Let $w(t)$ be the solution of equation (16.3) with initial data

$$w(0) = u_0.$$

Then the following inequalities similar to (16.11), (16.12) are valid:

$$||w(t)||_4^2 + ||w_t(t)||_2^2 + \int_0^t ||w_{tt}||_1^2 \, d\tau \leq C, \qquad (17.7)$$

$$t^{m_j} ||p_j(t)||_5^2 + t^{m_j} ||(p_j(t))_t||_2^2 + \int_0^t \tau^{m_j} ||(p_j)_{tt}||_1^2 \, d\tau \leq C. \qquad (17.8)$$

Indeed, since $w(0) = u_0 \in H_4(\Omega)$ and $||f(u_0)||_2 \leq C$, from (16.22) it follows that $w_t(0) \in H_2(\Omega) \cap H_1^0(\Omega)$. Multiplying equation (16.23) by Δw_{tt} in L_2 and integrating with respect to t, we deduce

$$||\Delta w_t(t)||^2 + \int_0^t ||w_{tt}||_1^2 \, d\tau \leq C.$$

Hence the right-hand side h of equation (16.25) is bounded in $H_2(\Omega)$ and so

$$||w(t)||_4 \leq C, \quad t \in [0, T].$$

Multiplying equation (16.26) by $t\Delta w_{tt}$ in L_2 we get the estimate (17.8) for $j = 0$. For $j = 1, 2, ...$ the proof is entirely analogous.

The function v is the second order boundary layer function and in the neighbourhood of the boundary $\Gamma_0 = \partial\Omega \cap \{y = 0\}$ has the form

$$v = v_0 = \varepsilon^2 c_0(x, t) e^{-y/\varepsilon},$$

where

$$c_0 = -\Delta w|_{y=0} = -\frac{\partial^2 w}{\partial y^2}\,|_{y=0}$$

(since $w(t)$ satisfies (17.7)).

The asymptotic expansion of $u(t)$ now has the form

$$u = w + v - \varepsilon^2 c + r \tag{17.9}$$

and the initial data (16.14) for $r(t)$ look like

$$r|_{t=0} = v^0 - (v - \varepsilon^2 c)|_{t=0}.$$

From (17.6) it follows that

$$\|r(0)\| \le C\varepsilon^2, \quad \|r(0)\|_1 \le C\varepsilon^{3/2}.$$

Substituting expansion (17.9) into equation (16.1) we obtain equation (16.13) by r in L_2 and using the boundary conditions (17.1) for $r(t)$, similarly to the estimate of the remainder in Theorem 16.1 we deduce the estimate (17.2) and the following estimate:

$$\varepsilon^2 t^m \|r\|_2^2 + t^m \|r\|_1^2 + \int_0^t \tau^m \|\partial_t r\|^2 \, d\tau \le C\varepsilon^4. \tag{17.10}$$

Let us multiply equation (16.13) by $t^m \Delta r$ in L_2:

$$\frac{1}{2}\partial_t(t^m \|\nabla r\|^2) + \varepsilon^2 t^m \|\nabla \Delta r\|^2 + t^m \|\Delta r\|^2$$
$$\le t^m |(dr, \Delta r)| + t^m |(R, \Delta r)| + \frac{m}{2} t^{m-1} \|\nabla r\|^2.$$

In a similar manner as in §16 we can prove that

$$\|R\| \le C\varepsilon^2.$$

Moreover, the boundedness of $tu(t)$ and $w(t)$ in $H_2(\Omega) \subseteq C(\Omega)$ implies the boundedness of $t^m d(u, w)$ in $C(\Omega)$ for certain m. Hence

$$\int_0^t (\varepsilon^2 \tau^m \|r\|_3^2 + \tau^m \|r\|_2^2) \, d\tau \le C\varepsilon^4. \tag{17.11}$$

The estimate (17.3) follows from (17.10) and (17.11).

Thus the asymptotic expansions of the trajectories $u(t) = S_t(\varepsilon)u_0$ of the problems (16.1) - (16.2) and (16.1), (17.1) for $t \in [t_1, t_2]$, $0 < t_1 < t_2 < +\infty$, are analogous; moreover, for the second problem the differential properties of the terms of its asymptotics are better than for the first boundary value problem.

Proposition 16.2 is also valid for the problem (16.1), (17.1), and the neighbourhood $\mathcal{O}_\delta(\mathcal{M})$ of the set $\mathcal{M}$ of the equilibrium points may be a set in $H_2(\Omega)$. Thus we establish the following result, similar to Theorem 16.1.

Theorem 17.2.

Let $\delta > 0$ and the trajectory $u^\varepsilon(t)$ of the semigroup $\{S_t(\varepsilon)\}$ of the problem (16.1), (17.1) belong to the set $U(R)$. Let for $t \in [t_1, t_2]$ the trajectory $u^\varepsilon(t)$ be outside a δ-neighbourhood $\mathcal{O}_\delta(\mathcal{M})$ of the set $\mathcal{M}$ of the equilibrium points of the semigroup $\{S_t(0)\}$ in $H_2(\Omega)$. Then

(1) $t_2 - t_1 \leq T$ and $T = T(\delta, R)$ doesn't depend on $\varepsilon \in (0, \varepsilon_0]$, $u^\varepsilon(t) \in U(R)$, t_1 or t_2;

(2) for $t \in (t_1, t_2)$ the asymptotic expansion of $u^\varepsilon(t)$ is valid, in which $w(t)$ is the trajectory of the semigroup $\{S_t(0)\}$, $v(t)$ is the second order boundary layer function such that

$$\|v\|_k \leq C_k \varepsilon^{5/2-k}, \quad k = 0, 1, 2, \ldots \qquad (17.12)$$

and the remainder satisfies the following estimate:

$$\varepsilon^2 \|r(t)\|_2^2 + \|r(t)\|_1^2 + \int_{t_1}^t (\varepsilon^2 \|r\|_3^2 + \|r\|_2^2 + \|\partial_t r\|^2)\, d\tau \leq C\varepsilon^4, \quad (17.13)$$

where $C = C(\delta, R)$ doesn't depend on ε or on $u^\varepsilon(t) \in U(R)$.

The proof of these assertions is similar to that of Theorem 16.2.

This theorem enables us to construct the global approximation of the trajectories of the dynamical system (16.1), (17.1). Let $U(R)$ be the set of solutions of (16.1), (17.1) which is determined in (16.4). Let $u(t)$ belong to $U(R)$ and $u(0) \notin \mathcal{O}_\delta(\mathcal{M})$ ($\mathcal{O}_\delta(\mathcal{M})$ is a neighbourhood of $\mathcal{M}$ in H_2). Denote by t_1^- the moment when $u(t)$ gets into $\mathcal{O}_\delta(\mathcal{M}) = [\mathcal{O}_\delta(\mathcal{M})]$ for the first time. We note that $t_1^- \leq T = T(\delta, R)$. As above we suppose that the set $\mathcal{M}$ consists of a finite number of equilibrium points, $\mathcal{M} = \{z_1, \ldots, z_N\}$, and every point is hyperbolic. Assume that $u(t_1^-) \in \mathcal{O}_\delta(z^1)$ where $z^1 = z_{k(1)}$. We denote by $\mathcal{O}_{\delta_0}^0(z^1)$ a neighbourhood of z^1 in $L_2(\Omega)$ and put δ_0 sufficiently small and δ sufficiently small so that $\mathcal{O}_\delta(z^1) \subseteq \mathcal{O}_{\delta_0}^0(z^1)$. Then as has been shown in Example 7.4, all the conditions of Theorem 6.1 are fulfilled for the problem (16.1), (17.1) ($\lambda = \varepsilon$), so for $t \geq t_1^-$ there exist stabilized asymptotics of $u(t)$ (see Example 7.4). Hence, denoting by t_1^+ the last moment of the stay of $u(t)$ in $\mathcal{O}_{\delta_0}^0(z^1)$ we have that there exists an approximation $\tilde{u}^1(t)$ of the part of $u(t)$, $t_1^- \leq t \leq t_1^+$, such that the following estimate is valid:

$$\|u(t) - \tilde{u}^1(t)\| \leq C\varepsilon^q, \quad t_1^- \leq t \leq t_1^+, \quad 0 < q < 1.$$

Recall that $\tilde{u}^1(t)$ consists of not more than two parts of trajectories of the limit equation (16.3):

$$\tilde{u}^1(t) = \left\{ \begin{array}{ll} \tilde{u}_1(t) = S_{t-t_1^-} u(t_1^-), & t_1^- \leq t \leq T, \\ \tilde{u}_2(t), & T < t \leq t_1^+, \; \tilde{u}_2(t) \in M^+(z^1), \end{array} \right\} \quad (17.14)$$

where $M^+(z^1)$ is the unstable manifold of the semigroup $\{S_t(0)\}$ passing through z^1.

In this way we determine the approximation $\tilde{u}^1(t)$ of the trajectory $u(t)$ for $t \in [t_1^-, t_1^+]$. Further, for $t \geq t_1^+$ an asymptotic expansion of the type (16.8) is valid:

$$u(t) = w_2(t) + v_2(t) + r_2(t), \quad t_1^+ \leq t \leq t_2^-,$$

where $w_2(t)$ is the solution of the limit problem (16.3) being determined as in Theorem 16.1, and t_2^- is the moment when the trajectory $u(t)$ gets into $\mathcal{O}_\delta(\mathcal{M})$ for the first time after t_1^+. For instance, $u(t_2^-) \in \mathcal{O}_\delta(z^2)$ where $z^2 = z_{k(2)} \in \mathcal{M}$. (Since there exists a global Lyapunov function for the semigroup $\{S_t(\varepsilon)\}$, $u(t)$ can't return to $\mathcal{O}_\delta(z^1)$.) Further construction is analogous. Thus we establish the following theorem.

Theorem 17.3.
Let the set $\mathcal{M} = \{z_1, ..., z_N\}$ of equilibrium points of semigroup $\{S_t(0)\}$ corresponding to (16.3) be finite and every point $z \in \mathcal{M}$ be hyperbolic. Then for any trajectory $u(t) = u^\varepsilon(t) \in U(R)$ (or $u(t) \in \mathcal{A}^\varepsilon$) such that $u(0) \notin \mathcal{O}_\delta(\mathcal{M})$ ($\mathcal{O}_\delta(\mathcal{M})$ is a δ-neighbourhood of $\mathcal{M}$ in $H_2(\Omega)$, δ is fixed and sufficiently small) there exists its global approximation $\tilde{u}(t)$, $t \in [0, +\infty)$. Let $u(t)$ pass for $0 \leq t < +\infty$ successively $\mathcal{O}_\delta(z^1), ..., \mathcal{O}_\delta(z^m)$ ($m \leq N$), $z^i \in \mathcal{M}$ ($z^i = z_{k(i)}$) and $\mathcal{O}_\delta(z^i) \cap \mathcal{O}_\delta(z^j) = \emptyset$ for $i \neq j$. Let at the moments $t_1^-, ..., t_m^-$ $u(t)$ get into these neighbourhoods and at the moments $t_1^+, ..., t_m^+$ ($t_m^+ = +\infty$) leave the neighbourhoods (in L_2) $\mathcal{O}_{\delta_0}^0(z^i)$ (which have been described above). Then on the intervals $[t_i^+, t_{i+1}^-]$ ($i = 0, 1, ..., m-1$, $t_0^+ = 0$) an asymptotic expansion of type (16.8) holds

$$u(t) = \tilde{u}(t) + r_i(t) \equiv w_i(t) + v_i(t) + r_i(t),$$

the remainder $r_i(t)$ satisfies (17.13), and on the intervals (t_i^-, t_i^+) $(i = 1, ..., m - 1)$

$$u^i(t) = \tilde{u}(t) + r_i(t), \qquad (17.15)$$

where $\tilde{u}^i(t)$ is determined as in (17.14) and

$$\|r_i(t)\| = \|u(t) - \tilde{u}^i(t)\| \leq C\varepsilon^q, \quad t_i^- < t < t_i^+, \quad 0 < q < 1.$$

The proof of this theorem follows from the construction given above and from Theorems 17.2 and 6.1.

Remark 17.1.

We have determined an approximation $\tilde{u}(t)$ of the type (17.15) to the function $u(t)$ in neighbourhoods $\mathcal{O}_{\delta_0}^0(z^i)$ only in the weak metric $\|\cdot\|_{L_2}$. The next problem is to construct an approximation to $u(t)$ in neighbourhoods of the equilibrium points z^i in the strong metrics $\|\cdot\|_2$. This question is still open.

Remark 17.2.

Let $u(t) \to z^m$ as $t \to +\infty$, $t > t_{m-1}^-$ and z^m be a stable equilibrium point (for example, $f'(z^m) \geq -(\lambda_1 - a^2)$).

Then we may consider the function $\overline{u}(t) = S_{t-t_{m-1}^-}(0)u(t_{m-1}^-)$ as the approximation to $u(t)$ (in the weak norm $\|\cdot\|_{L_2}$) in the neighbourhood $\mathcal{O}_{\delta_0}^0(z^m)$. Indeed, subtracting equations (16.1) for $u(t)$ and (16.3) for $\overline{u}(t)$, multiplying the resulting equation by $u(t) - \overline{u}(t)$ in L_2 and using the estimate (15.4) we get that

$$\|u(t) - \overline{u}(t)\| = \mathcal{O}(\varepsilon), \quad t \geq t_{m-1}.$$

Appendix

Non-autonomous dynamical systems and their attractors

by V. V. Chepyzhov and M. I. Vishik

In the Appendix we consider a new approach to the investigation of attractors for nonlinear non-autonomous partial differential equations. The Appendix is organised as follows. We start in §A1 with the definition of a process $\{U(t,\tau) \mid t \geq \tau, \tau \in \mathbf{R}\}$ acting on a Banach space E. §A1 contains also some examples of non-autonomous equations and systems of mathematical physics, which generate processes.

In §A2 we introduce the major concepts of absorbing set, attracting set and attractor for a given process $\{U(t,\tau)\}$. This section also deals with a family of processes $\{U_f(t,\tau) \mid t \geq \tau, \tau \in \mathbf{R}\}$ depending on some functional parameter f called the time symbol of a process. A special translation property holds for such a family of processes. It is shown that the study of a uniform attractor of such a family of processes can be reduced to the investigation of some semigroup $\{S(t)\}$ acting on some expanded phase space. We prove the main theorem about the existence of a uniform attractor of a family of processes with almost periodic time symbols.

§A3 is aimed at applying the general results of §A2 to the equations and systems we have considered in §A1. Using the main theorem we establish the existence of a uniform attractor of a family of processes generated by the two-dimensional Navier-Stokes equations with almost periodic exterior forces.

We prove the analogous results for the reaction-diffusion system with quasi-periodic and almost periodic terms and the hyperbolic equation with dissipation and with almost periodic terms.

In §§A4 and A5 we prove the finiteness of Hausdorff dimension of attractors for the two-dimensional Navier-Stokes system and reaction-diffusion system with quasi-periodic terms.

§A6 contains estimates for the Hausdorff dimension of sections of attractors corresponding to the equations and systems with general almost periodic terms, which we have considered in §A1 and §A3. All these results agree with the corresponding bounds for the autonomous case.

§A1 Processes corresponding to non-autonomous equations and systems in mathematical physics

Let E be a Banach space. We consider in E the general non-autonomous Cauchy problem:

$$\partial_t u = A(u,t), \quad t \geq \tau, \quad u \mid_{t=\tau} = u_\tau, \quad u_\tau \in E, \quad \tau \in \mathbf{R}, \qquad (A1.1)$$

where $A(u,t) : D \times \mathbf{R} \to H$, D and H are Banach spaces. For fixed t the operator $A(u,t)$ is a nonlinear operator defined on the domain D. Usually the set D is dense in H. We don't discuss here in what sense the function $u(t)$ is the solution of the problem (A1.1). We'll clarify it in each particular case. We assume that the initial value problem (A1.1) has a unique solution $u(t) \in E, \forall t > \tau$ for any $\tau \in \mathbf{R}$, and $u_\tau \in E$. We consider the two-parameter family of maps $\{U(t,\tau)\}$, $U(t,\tau) : E \to E$, parameters $\tau \in \mathbf{R}$, $t \geq \tau$:

$$U(t,\tau)u_\tau = u(t) \quad (u_\tau = u(\tau)), \qquad (A1.2)$$

where $u(t)$ is the solution of the equation (A1.1) with the initial data u_τ. We note that the family $\{U(t,\tau)\}$ has the following properties:
(1) $U(\tau,\tau) = I = $ identity;
(2) $U(t,s) \circ U(s,\tau) = U(t,\tau) \ \forall t \geq s \geq \tau \in \mathbf{R}$.

The first property is evident. The second property follows from the uniqueness of the solutions for the Cauchy problem (A1.1). Using (1) and (2) it is possible to define the important notion of a process acting on the space E. Let $\{U(t,\tau)\}$ be any two-parameter family of maps, $U(t,\tau) : E \to E$, $t \geq \tau$, $\tau \in \mathbf{R}$.

Definition A1.1.
A family of maps $\{U(t,\tau)\}$ is called a process on E if (1) and (2) hold. Thus the problem (A1.1) and the equality (A1.2) give an example of a process on E. Our definition of a process is equivalent to that of Dafermos [2], [3] and Haraux [3], but we use a slightly different notation.

A process is a generalization of a semigroup notion that usually associates with autonomous initial-value problems. In autonomous equations the nonlinear operator A in (A1.1) doesn't depend on t.

Now we consider some examples of equations and systems leading to the investigations of processes.

Example A1.1.   Two-dimensional Navier-Stokes system with time-dependent exterior force

Let Ω be an open bounded set of $\mathbf{R}^2$ with boundary $\partial\Omega$. We consider the system

$$\partial_t u + u_1 \partial_1 u_1 + u_2 \partial_2 u_2 - \nu \Delta u + \nabla p = \phi_0(t,x), \quad (\nabla, u) = 0. \quad (A1.3)$$

Here $x = (x_1, x_2) \in \Omega \in \mathbf{R}^2$,

$$u = u(x,t) = (u_1(x,t), u_2(x,t))$$

and $\phi_0(x,t) = (\phi_{01}(x,t), \phi_{02}(x,t))$ is a bounded function with values in $(L_2(\Omega))^2$, $t \in \mathbf{R}$.

The system (A1.3) is supplemented with the nonslip boundary condition

$$u\,|_{\partial\Omega} = 0. \quad (A1.4)$$

Let H (respectively H_1) be the closure of the set

$$V_0 = \{v \mid v \in (C_0^\infty(\Omega))^2, \ (\nabla, v) = 0\}$$

in the norm $\|\cdot\|$ (resp., $\|\cdot\|_1$) of the space $(L_2(\Omega))^2$ (resp., $(H_1(\Omega))^2$).

We denote by Π the orthogonal projection in $(L_2(\Omega))^2$ on the space H. We apply the projection Π to both sides of (A1.3) and have (see Ladyzhenskaya [1], Lions [1], Temam [1], Vishik & Fursikov [1])

$$\partial_t u + Lu + B(u,u) = \phi, \quad L = -\nu\Pi\Delta,$$

$$B(u,u) = \Pi \sum_{i=1}^{2} u_i \partial_i u, \quad \phi = \Pi\phi_0. \quad (A1.5)$$

The domain of L is $D(L) = H_1 \cap (H_2(\Omega))^2$, L is a self-adjoint positive operator in H.

The function $\phi(t) = \phi(t,x) = \Pi\phi_0(t,x)$ belongs to $L_\infty(\mathbf{R}, H)$.

We add to the equation (A1.5) an initial condition:

$$u\,|_{t=\tau} = u_\tau, \quad u_\tau \in H. \quad (A1.6)$$

For any $u_\tau \in H$ there exists a unique solution $u(t)$ of (A1.5), (A1.6) such that

$$u(t) \in C([\tau, +\infty), H) \cap L_2([\tau, T], H_1), \quad \forall T > \tau,$$
$$\partial_t u(t) \in L_2([\tau, T], H_{-1}) \quad \forall T > \tau,$$

where $H_{-1} = (H_1)^*$. The following inequalities are valid:

$$\|u(t)\|^2 \leq \|u(\tau)\|^2 \cdot \exp\{-\gamma_1 \nu \cdot (t - \tau)\} \tag{A1.7}$$
$$+ \gamma_1^{-1} \nu^{-2}(1 - \exp\{-\nu \gamma_1(t - \tau)\})\|\phi\|_{L_\infty(\mathbf{R}, H_1)}^2, \quad \gamma_1 > 0,$$

$$\|u(T)\|^2 + \nu \int_\tau^T \|u(\tau_1)\|_1^2 \, d\tau_1$$
$$\leq (T - \tau)\nu^{-1}\|\phi\|_{L_\infty(\mathbf{R}, H_{-1})}^2 + \|u(\tau)\|^2. \tag{A1.8}$$

The estimates (A1.7), (A1.8) can be deduced by taking the scalar product of (A1.5) with u, using properties of the bilinear operator $B(u, u)$ and applying the Gronwall Lemma (see Babin & Vishik [1], Temam [2]).

As a consequence we can determine the process $\{U(t, \tau)\}$ on H:

$$U(t, \tau)u_\tau = u(t),$$

where $u(t)$ is the solution of (A1.5), (A1.6) with initial data u_τ.

Example A1.2.   Time-dependent reaction-diffusion system

We denote by Ω an open bounded set of $\mathbf{R}^n$. We consider a boundary value problem involving a vector function $u = (u_1, u_2, ..., u_N)$ from $\Omega \times \mathbf{R}_\tau$ into $\mathbf{R}^N$ ($\mathbf{R}_\tau = [\tau, +\infty)$); u satisfies the following system:

$$\partial_t u = a\Delta u - f(u, x, t) - \phi(x, t),$$
$$u|_{\partial\Omega} = 0 \quad (\text{or } \frac{\partial u}{\partial \nu}|_{\partial\Omega} = 0), \qquad x \in \Omega \in \mathbf{R}^n, \tag{A1.9}$$

where $a + a^*$ is a positive matrix, $f = (f_1, ..., f_N)$, $\phi = (\phi_1, ..., \phi_N)$. We assume that f, f_u', f_x', $f_t' \in C(\mathbf{R}^N \times \Omega \times \mathbf{R}, \mathbf{R}^N)$, $\phi(x, t) \in C_b(\mathbf{R}, (L_2(\Omega))^N)$. Let f satisfy the conditions

$$\gamma_2|u|^p - C_o \leq (f(u, x, t), u) \leq \gamma_1|u|^p + C_0',$$
$$\gamma_i > 0, \quad 2 \leq p \leq \frac{2n}{(n-2)}, \quad n \geq 3, \quad f(0, x, t) + 0 \tag{A1.10}$$
$$(f_u'(u, x, t)v, v) \geq -C|v|^2 \quad \forall v \in \mathbf{R}^N, \tag{A1.11}$$
$$|f(u, x, t)| \leq C(|u|^{(p-1)}), \quad |f_u'(u, x, t)| \leq C(1 + |u|), \tag{A1.12}$$
$$|f_u'(u, x, t)| \leq C(1 + |u|)^{p-2)} \tag{A1.13}$$

for any $u \in \mathbf{R}^N$, $x \in \overline{\Omega}$, $t \in \mathbf{R}$. *Conditions (A1.10) - (A1.13) can be generalized. We use them only for simplicity.*

We complete the system (A1.9) with an initial condition:

$$u|_{t=\tau} = u_\tau, \quad u_\tau \in (L_2(\Omega))^N. \tag{A1.14}$$

For u_τ given in $(L_2(\Omega))^N$ there exists a unique solution u of (A1.9), (A1.14) such that

$$u(t) \in L_\infty(\mathbf{R}_\tau, (L_2(\Omega))^N) \cap L_2([\tau, T], (H_1^0(\Omega))^N),$$
$$\partial_t u(t) \in L_2([\tau, T], (H_{-1}(\Omega))^N) \quad \forall T \geq \tau.$$

We write $\mathbf{R}_\tau = [\tau, +\infty)$.

Finally the following estimates hold:

$$\|u(t)\|^2 \leq \exp(-\gamma^2(t-\tau)) \cdot \|u(\tau)\|^2$$
$$+ \frac{1}{\gamma^2}\left(C_1 + \sup_{t\in\mathbf{R}}\|\phi(t,\cdot)\|^2_{-1}\right) \quad \forall t \geq \tau, \tag{A1.15}$$

$$\|u(t)\|^2 + \gamma_1^2 \int_\tau^t \|u(\tau_1)\|_1^2 \, d\tau_1$$
$$\leq C_2(t-\tau) + \|u(\tau)\|^2 + \int_\tau^t \|\phi(\tau_1,\cdot)\|^2_{-1} \, d\tau_1, \tag{A1.16}$$

$$(t-\tau_1)\|u(t)\|_1^2$$
$$\leq C_3 \int_{\tau_1}^t \left[(\tau_2-\tau_1)\|\phi(\tau_2)\|^2 + \|u(\tau_2)\|_1^2(\tau_2-\tau_1+1) + 1\right] d\tau_2. \tag{A1.17}$$

Constants γ, γ_1, C_1, C_2, and C_3 depend only on Ω, a, C_0 and C.

To prove (A1.15) one has to multiply the system (A1.9) by $u(t)$, to integrate and to use inequalities (A1.10) - (A1.13). One has

$$\partial_t\|u\|^2 + \gamma_1^2\|u\|_1^2 \leq C_1 + \|\phi(t,\cdot)\|^2_{-1}. \tag{A1.18}$$

There exists a constant γ^2 such that

$$\gamma_1^2\|u\|_1^2 \geq \gamma^2\|u\|^2. \tag{A1.19}$$

Using the Gronwall Lemma we infer from (A1.18), (A1.19) the inequality (A1.15). The estimate (A1.16) follows directly from (A1.15) by integration. To show (A1.17) we take the scalar product of (A1.9) with

$-(t - \tau_1)\Delta u(t)$. According to standard procedures, the estimate (A1.17) can be derived from (A1.10) - (A1.13).

The problem (A1.9), (A1.14) generates the process $\{U(t, \tau)\}$ which acts on the space $E = (L_2(\Omega))^N$, $U(t, \tau) : E \to E$, $U(t, \tau)u_\tau = u(t)$, $t \geq \tau$, where $u(t)$ is a solution of (1.9), (1.14).

Example A1.3.   Time-dependent hyperbolic equation with dissipation

Let Ω be an open bounded set of $\mathbf{R}^3$ (for simplicity) with a boundary $\partial\Omega$ sufficiently regular. The unknown function $u = u(x, t)$, $x \in \Omega$, $t \in \mathbf{R}_\tau$, satisfies the equation

$$\partial_t^2 u + \gamma \partial_t u = \Delta u - f(u, t) - \phi(x, t), \quad u|_{\partial\Omega} = 0 \qquad (A1.20)$$

$(f(u, t)$ can be replaced by $f(u, x, t))$. We suppose that

$$f(u, t) \in C^2(\overline{\Omega}, \mathbf{R})$$

and

$$\mathcal{F}(u, t) = \int_0^u f(v, t)\, dv, \quad \mathcal{F}(u, t) \geq -mu^2 - c_m,$$

where m is sufficiently small;

$$\left. \begin{aligned} uf(u, t) - c\mathcal{F}(u, t) + mu^2 &\geq -c_m, \\ |f_u'(u, t)| &\leq C(1 + |u|)^\rho, \\ |\partial_t f(u, t)| &\leq C(1 + |u|)^{\rho+3}, \\ \partial_t \mathcal{F}(u, t) &\leq \delta^2 \mathcal{F}(u, t) + C \end{aligned} \right\} \qquad (A1.21)$$

$\forall (u, t) \in \mathbf{R} \times \mathbf{R}$, where δ^2 is sufficiently small.

The case of limit exponent $\rho = 2$ for the autonomous equation (A1.20) has been considered by Babin & Vishik [1], [9], Ladyzhenskaya [2] and other authors. The case $\rho < 2$ has been studied by Haraux [3], [4], Temam [2] and other authors. For brevity we shall discuss only the case $\rho < 2$.

We assume that $\phi(x, t) \in C_b(\mathbf{R}, L_2(\Omega))$. As usually we supplement the equation (A1.20) with initial conditions:

$$u|_{t=\tau} = u_\tau(x), \quad \partial_t u|_{t=\tau} = p_\tau(x). \qquad (A1.22)$$

We consider the following Hilbert space of vector functions:

$$E = \{y(x) = (u(x), p(x)) \mid u \in H_1 = H_1^0(\Omega), p \in H = L_2(\Omega)\}$$

with the norm $||y||^2 = ||u||_1^2 + ||p||^2$. We'll briefly describe the main properties of the Cauchy problem (A1.20), (A1.22). The detailed proof of these results in the autonomous case is given in Babin & Vishik [1], Temam [2]. The same arguments based on (A1.21) can be used in the non-autonomous case.

Let $(u_\tau(x), p_\tau(x)) = y_\tau(x) = y_\tau \in E$; then the Cauchy problem (A1.20), (A1.22) has a unique solution $u(x,t) = u(t)$ and

$$y(t) = (u(t), \partial_t u(t)) \in C_b(\mathbf{R}_\tau, E),$$

where $\mathbf{R}_\tau = [\tau, +\infty)$, C_b is a space of bounded continuous functions. This statement allows us to define the operators in E

$$U(t, \tau) : y_\tau \to y(t).$$

These operators have the process properties.

§A2 Families of processes and their attractors: the general framework

For any Banach space E we denote by $\mathcal{B}(E)$ the set of all bounded subsets of E. Let $\{U(t, \tau)\}$ be a process on E. As in the theory of semi-groups we introduce the concepts of an absorbing set and an attracting set.

Definition A2.1.
The set $B_1 \subseteq E$ is called absorbing if for any $\tau \in \mathbf{R}$ and any $B \in \mathcal{B}(E)$ there exists $T = T(\tau, B)$ such that for $t \geq T$ the inclusion $U(t, \tau)B \subseteq B_1$ is valid.

A process which has a compact absorbing set will be called compact.

Definition A2.2.
The set $K \subseteq E$ is called attracting if for any $\tau \in \mathbf{R}$ and any $B \in \mathcal{B}(E)$ the following equality holds:

$$\lim_{t \to +\infty} \mathrm{dist}_E(U(t, \tau)B, K) = 0.$$

For arbitrary $X \subset E$, $Y \subset E$ we set

$$\mathrm{dist}_E(X, Y) = \sup_{x \in X} \inf_{y \in Y} ||x - y||_E.$$

A process having a compact attracting set is called asymptotically compact. (See Haraux [3]). It is evident that an absorbing set is an attracting set and a compact process is asymptotically compact. In general the converse statements are not true.

In the theory of semigroups the attractor is the natural mathematical object describing the behaviour of trajectories of a semigroup when $t \to +\infty$. The global or maximal attractor of a semigroup is a closed attracting set which is invariant for the semigroup.

To define the attractor of a process we have to use the concept of minimality instead of invariance.

Definition A2.3.
A closed set $\mathcal{A} \subset E$ is called the attractor of a process $\{U(t,\tau)\}$ if it is attracting and any closed attracting set of the process $\{U(t,\tau)\}$ contains $\mathcal{A}$. (See Haraux [3].)

It is clear that a process can have at most one attractor.

We'll see below that it is much more useful to consider not one but a family of processes $\{U_f(t,\tau)\}$ depending on $f \in F$. The functional parameter f is called a symbol (or time symbol) of the process $\{U_f(t,\tau)\}$.

We are going to investigate properties of processes which are uniform with respect to $f \in F$. We modify Definitions A2.1, A2.2, A2.3 a little for a family of processes.

Let $\{U_f(t,\tau)\}$, $f \in F$, be a family of processes on E.

Definition A2.4.
The set $B_1 \subseteq E$ is called uniformly absorbing (with respect to F), if for any $\tau \in \mathbf{R}$ and any $B \in \mathcal{B}(E)$ there exists $T = T(\tau, B)$ such that the inclusion

$$\bigcup_{f \in F} U_f(t,\tau)B \subseteq B_1$$

is valid for $t \geq T$.

Definition A2.5.
The set $K \subseteq E$ is called uniformly attracting if

$$\sup_{f \in F} \operatorname{dist}_E(U_f(t,\tau)B, K) \to 0 \quad (t \to +\infty)$$

for any $\tau \in \mathbf{R}$ and $B \in \mathcal{B}(E)$.

A family of processes which has a compact uniformly absorbing set is called uniformly compact. A family of processes having a compact uniformly attracting set is called uniformly asymptotically compact. (See Haraux [3].)

Finally we determine a uniform attractor of a family of processes.

Definition A2.6.
A closed set $\mathcal{A}_F \subseteq E$ is called the uniform attractor of a family of processes $\{U_f(t, \tau)\}$, $f \in F$, if it is uniformly attracting (with respect to F) and each closed uniformly attracting set contains $\mathcal{A}_F$.

Definitions A2.1, A2.2 and A2.3 reduce to Definitions A2.4, A2.5 and A2.6 when the set of parameters F consists of one element. Note that for each fixed $f \in F$ the process $\{U_f(t, \tau)\}$ can have an attractor $\mathcal{A}_{\{f\}}$. It is evident that

$$\bigcup_{f \in F} \mathcal{A}_{\{f\}} \subseteq \mathcal{A}_F. \tag{A2.1}$$

At the same time the converse inclusion may not hold, i.e. in general a uniform attractor $\mathcal{A}_F$ of a family of processes $\{U_f(t, \tau)\}$, $f \in F$, is larger than the union of all $\mathcal{A}_{\{f\}}$. The counterexample is given by Haraux [3].

In the present work we investigate just uniform attractors, determine their structure and estimate the upper bound of their dimension. Owing to (A2.1) one can also establish certain facts concerning the non-uniform attractors.

We shall work only with continuous families of processes.

Definition A2.7.
A family of processes $\{U(t, \tau)\}$, $f \in F$, acting on E is called $(E \times F, E)$-continuous, if for any $\tau \in \mathbf{R}$, $t \geq \tau$ the map

$$U_f(t, \tau)u : E \times F \to E$$

is continuous with respect to $u \in E$, $f \in F$.

We consider the following particular case of the Cauchy problem (A1.1):

$$\begin{aligned} \partial_t u(t) &= -A(u(t), f(t)), \quad t \geq \tau \\ u|_{t=\tau} &= u_\tau \in E, \quad \tau \in \mathbf{R}, \end{aligned} \tag{A2.2}$$

where $A : D \times M \to E$, M is a metric space, $f(t)$ is a function with values in M, $t \in \mathbf{R}$. As for problem (A1.1) we assume that the initial value problem (A2.2) is well posed and that it generates the process $\{U_f(t, \tau)\}$ on E. It is clear that $f = f(t)$ is a natural parameter of the problem (A2.2) and of the process $\{U_f(t, \tau)\}$.

$f(t)$ is called the time symbol of the process $\{U_f(t,\tau)\}$. As a set F of symbols we consider the set of all translations of the function $f(t)$:

$$F = \{f^h(t)\}_{h\in\mathbf{R}}, \quad \text{where } f^h(t) = f(t+h).$$

Each symbol $f^h(t)$ generates a process $\{U_{f^h}(t,\tau)\}$ by the following problem:

$$\partial_t u(t) = -A(u(t), f^h(t)), \quad t \geq \tau, \quad u|_{t=\tau} = u_\tau. \quad (A2.3)$$

Obviously the process $\{U_{f^h}(t,\tau)\}$ can be obtained from the process $\{U_f(t,\tau)\}$ by translation on h:

$$U_{f^h}(t,\tau)u_\tau = U_f(t+h, \tau+h)u_\tau. \quad (A2.4)$$

To show (A2.4) it is sufficient to replace time $t+h$ by t' in (A2.3). The equation (A2.4) is called the translation equality. The following reasoning is the key idea. We'll construct from the family of processes $\{U_{f^h}(t,\tau)\}$, $h \in \mathbf{R}$, the semigroup acting on the set $E \times \{f^h\}_{h\in\mathbf{R}}$.

We determine the family of operators $\{S(t)\}$:

$$S(t_1)(u_0, f^h) = (U_{f^h}(t_1,0)u_0, f^{h+t_1}(\cdot)), \quad t_1 \geq 0. \quad (A2.5)$$

We let (u_0, f^h) denote an arbitrary element from $E \times \{f^h\}_{h\in\mathbf{R}}$.

Proposition A2.1.
   The family $\{S(t)\}_{t\geq 0}$ is a semigroup.

Proof   Evidently, $S(0) = I = $ identity. Let's check the semigroup property

$$S(t_1 + t_2) = S(t_1) \cdot S(t_2).$$

Indeed,

$$\begin{aligned}
S(t_1 + t_2)(u_0, f^h) &= (U_{f^h}(t_1 + t_2, 0)u_0, f^{h+t_1+t_2}) \\
&= (U_{f^h}(t_1 + t_2, t_1) \circ U_{f^h}(t_1, 0)u_0, f^{(h+t_1)+t_2}) \\
&= (U_{f^{h+t_1}}(t_2, 0) \circ U_{f^h}(t_1, 0)u_0, f^{(h+t_1)+t_2}) \\
&= S(t_2)(U_{f^h}(t_1, 0)u_0, f^{h+t_1}) \\
&= S(t_2) \circ S(t_1)(u_0, f^h).
\end{aligned}$$

In this chain of equalities we have used property (2) of a process and the translation equality (A2.4).

The semigroup $\{S(t)\}$ acts on the set $E \times \{f^h\}$, but this set in general is not closed because the second component $\{f^h\}$ in general is not closed. Thus we have to close the set $\{f^h\}$ in a proper metric space and expand the semigroup to this set. We consider the space $C_b(M) = C_b(\mathbf{R}, M)$ of continuous bounded functions $g(t)$, $t \in \mathbf{R}$, with values in a complete metric space M with metric

$$\rho(f_1(\cdot), f_2(\cdot))_{C_b(M)} = \sup_{t \in \mathbf{R}} \rho_M(f_1(t), f_2(t)), \qquad (A2.6)$$

where ρ_M is the metric on M.

Definition A2.8.

A function $f(t) \in C_b(M)$ is called almost periodic if the set $\{f^h(t)\}_{h \in \mathbf{R}}$ is precompact in $C_b(M)$. (See Levitan & Zhikov [1]).

The set $H(f) = \overline{\{f^h\}_{h \in \mathbf{R}}}$ is called the hull of the function f. (The over-lining means closure in $C_b(M)$). Note that $g(t) \in H(f)$ if there exists a sequence $\{h_n\} \subset \mathbf{R}$ such that

$$\rho(h^{h_n}(\cdot), g(\cdot))_{C_b(M)} \to 0 \quad \text{when } n \to +\infty.$$

We suppose that the function $f(t)$ in the problem (A2.5) is almost-periodic. Let $H(f)$ be the hull of $f(t)$. We consider the family of Cauchy problems

$$\begin{aligned} \partial_t u(t) &= -A(u(t), g(t)), \quad t \geq \tau, \quad g(t) \in H(f), \\ u|_{t=\tau} &= u_\tau \in E, \quad \tau \in \mathbf{R}. \end{aligned} \qquad (A2.7)$$

We suppose that the problem (A2.7) is well posed for each $g \in H(f)$, $\tau \in \mathbf{R}$ and $u_\tau \in E$. Therefore one can construct the family of processes $\{U_g(t, \tau)\}$, $g \in H(f)$, acting on E.

It follows from (A2.4) that the family $\{U_g(t, \tau)\}$, $g \in H(f)$, satisfies the translation equality

$$U_{g^h}(t, \tau)u = U_g(t + h, \tau + h)u. \qquad (A2.8)$$

As for the problem (A2.5) we determine the family of operators $\{S(t)\}$, $t \geq 0$, acting on the set $E \times H(f)$:

$$S(t_1)(u_0, g) = (U_g(t_1, 0)u_0, g^{t_1}(\cdot)) = (u(t_1), g^{t_1}), \qquad (A2.9)$$

where $u(t)$ is the solution of the problem (A2.7) with the symbol g and initial data $u(0) = u_0$.

Proposition A2.2.
The family $\{S(t)\}$, $t \geq 0$, is a semigroup on $E \times H(f)$.

The proof is the same as the proof of Proposition A2.1.

A solution $u(t)$ of (A2.7) determined on $t \in \mathbf{R}$ is called a bounded complete trajectory if the set $\{u(t) \mid t \in \mathbf{R}\}$ is bounded in E.

Now everything is ready to formulate the main theorem about attractors of the semigroup $\{S(t)\}$ and the family $\{U_g(t,\tau)\}$, $g \in H(f)$.

Theorem A2.1.
Let $f(t)$ be an almost-periodic function, the family of processes

$$\{U_g(t,\tau)\}, \quad g \in H(f),$$

corresponding to the family of Cauchy problems (A2.7) be uniformly asymptotically compact and $(E \times H(f), E)$-continuous. Then the semigroup $\{S(t)\}$ acting on $E \times H(f)$ according to (A2.9) possesses the compact attractor $\mathcal{A}$. The set $\mathcal{A}$ is invariant with respect to $\{S(t)\}$: $S(t)\mathcal{A} = \mathcal{A}$ $\forall t \geq 0$. Moreover

$$\mathcal{A} = \{u(0), g\}; \tag{A2.10}$$

$u(t)$ is any bounded complete trajectory of the problem (A2.7) with arbitrary $g \in H(f)$.

Proof   Let K be a compact uniformly attracting set of the family $\{U_g(t,\tau)\}$, $g \in H(f)$. Due to (A2.9) the set $K \times H(f)$ is also attracting for the semigroup $\{S(t)\}$. The continuity of the family $\{U_g(t,\tau)\}$ implies that the semigroup $\{S(t)\}$ is continuous. By the general theorem about the existence of an attractor of a semigroup there exists a compact invariant attractor $\mathcal{A}$ of the semigroup $\{S(t)\}$: $S(t)\mathcal{A} = \mathcal{A}$ $\forall t \geq 0$. (See Babin & Vishik [1], Temam [2]).

Let's check (A2.10). Let $(u, g) \in \mathcal{A}$. Due to the invariance property there exists a bounded trajectory $(u(s), g^s)$, $s \in \mathbf{R}$, such that $u(0) = u$, $g^0 = g$ and the following equality holds:

$$S(t_1)(u(s), g^s) = (u(s + t_1), g^{s+t_1}(\cdot)).$$

By (A2.9) we get

$$U_{g^s}(t, 0)u(s) = u(s + t). \tag{A2.11}$$

The translation equality (A2.8) implies

$$U_{g^s}(t,0) = U_{g^0}(t+s,s),$$

i.e. by (A2.11)

$$U_g(t+s,s)u(s) = u(s+t) \quad \forall s \in \mathbf{R}, t \geq 0.$$

Finally, changing notation $(t+s \mapsto t, s \mapsto \tau)$ we deduce

$$U_g(t,\tau)u(\tau) = u(t) \quad \forall t \geq \tau, \tau \in \mathbf{R}.$$

Hence $u(t)$ is the bounded complete trajectory of the problem (A2.7) with time symbol g.

Using similar reasoning one can prove that any bounded complete trajectory $u(t)$ of the problem (A2.7) with a symbol g generates a bounded trajectory $(u(s), g^s)$, $s \in \mathbf{R}$, of the semigroup $\{S(t)\}$. Thus $(u(0), g) \in \mathcal{A}$. The theorem is proved.

Now we establish a relation between the attractor $\mathcal{A}$ of the semigroup $\{S(t)\}$ and the uniform attractor of the family $\{U_g(t,\tau)\}$, $g \in H(f)$. We consider two projections π_1 and π_2 in $E \times H(f)$:

$$\pi_1 : E \times H(f) \to E, \qquad \pi_1(u,g) = u,$$
$$\pi_2 : E \times H(f) \to H(f), \qquad \pi_2(u,g) = g.$$

Corollary A2.1.
(1) $\pi_1(\mathcal{A}) = \mathcal{A}_{H(f)}$ *is the uniform attractor of the family of processes* $\{U_g(t,\tau)\}$, $g \in H(f)$.
(2) *(Haraux [3])*

$$\mathcal{A}_{H(f)} = \{u(0) \,|\, u(t) \text{ is any bounded complete trajectory of prob-}$$
$$\text{lem (A2.7) with an arbitrary symbol } g \in H(f)\}$$

(3) $\pi_2(\mathcal{A}) = H(f)$.

Proof   Evidently the set $\pi_1(\mathcal{A})$ is uniformly attracting for the family $\{U_g(t,\tau)\}$, $g \in H(f)$.

Due to (A2.10) we have $\pi_1(\mathcal{A}) = \{u(0) \,|\, u(t)$ is any bounded complete trajectory of the problem (A2.7) with any symbol $g \in H(f)\}$. Note that any closed uniformly attracting set of the family $\{U_g(t,\tau)\}$, $g \in H(f)$, contains the set of all bounded trajectories determined by the right-hand side of the above equality. Hence the minimal property is valid and we have $\mathcal{A}_F = \pi_1(\mathcal{A})$.

Property (3) follows from the equality

$$\pi_2(S_t(K \times H(f))) = H(f) \quad \forall t \geq 0,$$

where K is the uniformly attracting set of the family $\{U_g(t, \tau)\}$, $g \in H(f)$.

Corollary A2.2.
For any $g \in H(f)$ there exists a bounded complete trajectory $u(t)$, $t \in \mathbf{R}$, of the problem (A2.7) with time symbol g. This follows from property (3) of Corollary A2.1.

In the next section we consider some examples of equations and systems in the form (2.2) and (2.7). Also, in §A6 we shall briefly describe some more general equations which have the symbol $f = f(u, t)$ dependent on u with values in more complicated spaces. Theorem A2.1 is still valid for these problems.

§A3 Attractors of families of processes generated by non-autonomous equations and systems in mathematical physics

Example A3.1. Navier-Stokes equations $(n = 2)$ with almost periodic exterior forces
We assume that in Example A1.1 the function $\phi(x, t)$ is almost periodic with values in H. Let $H(\phi)$ be a hull of ϕ. We consider the family of Cauchy problems:

$$\partial_t u + Lu + B(u, u) = g(x, t), \quad g \in H(\phi), \quad t \geq \tau,$$
$$u|_{t=\tau} = u_\tau \in H. \tag{A3.1}$$

Obviously for any $g \in H(\phi)$ the problem has a unique solution $u(t)$ and the inequalities (A1.7), (A1.8) hold for $u(t)$, because

$$\|\phi\|_{L_\infty(\mathbf{R}, H_{-1})} = \|g\|_{L_\infty(\mathbf{R}, H_{-1})}.$$

Thus the family of problems (A3.1) generates the family of processes $\{U_g(t, \tau)\}$, $g \in H(\phi)$, acting on the space H. This family satisfies the following properties.

Lemma 3.1.
(1) The family $\{U_g(t,\tau)\}$, $g \in H(\phi)$ is uniformly bounded on H, i.e.

$$\|U_g(t,\tau)u_\tau\| \leq R_0(\|u_\tau\|) \quad \forall t \geq \tau, \tau \in \mathbf{R}.$$

(2) If $t > \tau$, then $U_g(t,\tau)u_\tau \in H_1 \;\forall g \in H(f)$. Moreover, if $\tau + \delta < t \leq \tau + 1$ ($\delta < 1$), then

$$\|U_g(t,\tau)u_\tau\|_1 \leq R(\|u_\tau\|, \delta) \quad \forall g \in H(f), \forall \tau \in \mathbf{R}.$$

Proof   Property (1) follows from the bound (A1.7). Property (2) can be proved by taking the scalar product of equation (A3.1) with $(t-\tau)Lu$. (See Babin & Vishik [1], Temam [2].)

Proposition A3.1.
The family $\{U_g(t,\tau)\}$, $g \in H(f)$ is uniformly compact and $(H \times H(\phi), H)$-continuous.

Proof   The estimate (1.7) implies that the set

$$B_0 = \{u \in H \mid \|u\|^2 \leq 2R_0^2\}, \tag{A3.2}$$

where $R_0^2 = 2\nu^{-2}\gamma_1^{-1} \cdot \|\phi\|_{L_\infty(\mathbf{R},H_{-1})}^2$, is uniformly absorbing for the family $\{U_g(t,\tau)\}$, $g \in H(f)$. Due to property (2) of Lemma A3.1 the set $B_1 = \{\|u\|_1 \leq R_1\} \subset H_1$, $R_1 = R(R_0, \delta)$, is also uniformly absorbing. The set B_1 is compact in H, therefore the family $\{U_g(t,\tau)\}$, $g \in H(f)$ is uniformly compact.

Let's show that the family $\{U_g(t,\tau)\}$, $g \in H(f)$ is $(H \times H(\phi), H)$-continuous. For simplicity let $\nu = 1$. Let $u_n(\tau) \to u(\tau)$ in H, $g_n(s) \to g(s)$ in $C_b(\mathbf{R}, H)$. We write

$$u_n(t) = U_{g_n}(t,\tau)u_n(\tau), \quad u(t) = U_g(t,\tau)u(\tau).$$

We shall prove that $u_n(t) \to u(t)$ in H ($n \to +\infty$).

Subtracting the equations (A1.5) for u, $g = \phi$ and u_n, $g_n = \phi_n$ and taking the scalar product with $u - u_n$ we obtain the following equation for $w(t) = u(t) - u_n(t)$:

$$\frac{1}{2}\partial_t\|w(t)\|^2 + \|w(t)\|_1^2 + (B(w,u),w) + (B(u,w),w) - (B(w,w),w)$$
$$= (g - g_n, w).$$

It is well known that $\langle B(u,w), w \rangle = 0$, hence

$$\frac{1}{2}\partial_t \|w(t)\|^2 + \|w(t)\|_1^2 + \langle B(w,u), w \rangle \le M\|g - g_n\|_{-1}^2 + \frac{1}{4}\|w\|_1^2. \quad (A3.3)$$

We have

$$|\langle B(w,u), w \rangle| \le \|w\|_{0,4}^2 \cdot \|u\|_1 \le C_0 \|w\| \cdot \|w\|_1 \cdot \|u\|_1$$
$$\le \frac{1}{4}\|w(t)\|_1^2 + C\|u(t)\|_1^2 \cdot \|w(t)\|^2.$$

This and (A3.3) imply

$$\partial_t \|w(t)\|_1^2 \le C\|u(t)\|_1^2 \cdot \|w(t)\|^2 + M\|g(t) - g_n(t)\|_{-1}^2.$$

According to the Gronwall Lemma

$$\|w(t)\|^2 \le \|w(\tau)\|^2 \cdot \exp\left(C\int_\tau^t \|u(\tau_1)\|_1^2 \, d\tau\right)$$
$$+ M\int_\tau^t \exp\left(C\int_{\tau_1}^t \|u(\tau_2)\|_1^2 \, d\tau_2\right) \cdot \|g(\tau_1) - g_n(\tau_1)\|^2 \, d\tau_1. \quad (A3.4)$$

Due to (A1.8)

$$\int_\tau^t \|u(\tau_1)\|_1^2 \, d\tau_1 \le M_1(t) \text{ for } t \ge \tau,$$

$g_n(\tau_1) \to g(\tau_1)$ in $C_b([\tau, t], H)$ $(n \to +\infty)$, and

$$\|w(\tau)\|^2 = \|u(\tau) - u_n(\tau)\|^2 \to 0$$

$(n \to +\infty)$.

Finally, we infer from (A3.4) that

$$\|w(t)\|^2 = \|u(t) - u_n(t)\|^2 \to 0 \quad (n \to +\infty),$$

i.e. the family of processes $\{U_g(t,\tau)\}$, $g \in H(\phi)$, is $(H \times H(\phi), H)$-continuous.

Let $\{S(t)\}$ be a semigroup acting on $H \times H(\phi)$ and defined by formula (A2.9), i.e.

$$S(t_1)(u_0, g) = (U_g(t_1, 0)u_0, g^{t_1}(\cdot)),$$

where $g^{t_1}(t) = g(t + t_1)$.

The results of §A2 and Proposition A3.1 imply the following theorem.

Theorem A3.1.

(1) The set $\mathcal{A} = \{(u(0), g) \mid u(t)$ is any bounded complete trajectory of the problem (A3.1) for arbitrary $g \in H(\phi)\}$ is the attractor of the semigroup $\{S(t)\}$, acting on $H \times H(\phi)$.

(2) The set $\mathcal{A}_{H(\phi)} = \{u(0) \mid u(t)$ is any bounded complete trajectory of the problem (A3.1) with arbitrary symbol $g \in H(\phi)\}$ is the uniform attractor of the family $\{U_g(t, \tau)\}$, $g \in H(\phi)$, generated by the problem (A3.1).

Example A3.2.   Time-dependent reaction-diffusion system with almost periodic terms

We consider Example A1.2 from §A1. We assume that the function $\phi(x, t)$ is almost periodic with values in $(L_2(\Omega))^N$. The case of an almost periodic function $f(u, x, t)$ will be considered in §A6. In the current section we'll investigate for simplicity the case of a general quasi-periodic function f:

$$f(u, x, t) = F(u, x, \lambda_1 t, \lambda_2 t, ..., \lambda_k t), \qquad (A3.5)$$

where the function $F(u, x, \omega_1, ..., \omega_k)$ is periodic with period 2π with respect to each argument ω_j $(j = 1, ..., k)$,

$$F(u, x, \omega_1, ..., \omega_j + 2\pi, ..., \omega_k) = F(u, x, \omega_1, ..., \omega_j, ..., \omega_k),$$

$\lambda_1, ..., \lambda_k$ are positive numbers. We write $\Lambda = (\lambda_1, ..., \lambda_k)$. We assume that the numbers $\lambda_1, ..., \lambda_k$ are rationally independent. (Otherwise one can reduce the number of parameters ω_j.)

Let T^k be a k-dimensional torus. We consider a function $\omega_0(t)$, $t \in \mathbf{R}$, with values in T^k:

$$\omega_0(t) = (\lambda_1 t, \lambda_2 t, ..., \lambda_k t) \bmod 2\pi = \Lambda t \bmod 2\pi \equiv [\Lambda t].$$

Obviously $\omega_0(t)$ is an almost-periodic function with values in the metric space T^k with metric induced from $\mathbf{R}^{k+1}$. Using new notation we can write

$$f(u, x, t) = F(u, x, \omega_0(t)) = F(u, x, [\Lambda t]).$$

Hence a time symbol of a problem (A1.9) is a pair of functions $(\omega_0(t), \phi(t)) = P(t)$. It is clear that $P(t)$ is an almost periodic vector function with values in $T^k \times (L_2(\Omega))^N$.

Let $H(P)$ be a hull of this function. Obviously $H(P) \subseteq H(\omega_0(t)) \times H(\phi)$. In general the inverse embedding may not be true. Nevertheless we shall consider the family of Cauchy problems with symbols from $H(\omega_0(t)) \times H(\phi)$. This set we shall denote by $H(P)$ for brevity. At the same time all the results remain true for the real hull of the function P. Note that

$$H(\omega_0(t)) = \{[\Lambda t + \omega] \mid \omega \in T^k\}$$

because the numbers $\lambda_1, ..., \lambda_k$ are rationally independent. In fact the set $H(\omega_0(t))$ is isometrically equivalent to the set T^k. A function $\omega(t) = [\Lambda t + \omega]$ corresponds to $\omega \in T^k$. Using this equivalence we get

$$H(P) = T^k \times H(\phi), \quad P = (0, g(t)), \quad \omega = 0, \quad g(t) \in H(\phi(t)).$$

We let $G(t)$ denote any function from $H(P)$. We have

$$G(t) = (\omega, g(t)), \quad \omega \in T^k, \quad g(t) \in H(\phi).$$

Any translation $G^h(t)$ of a function $G(t)$ is equal to $G^h(t) = ([\omega + \Lambda h], g(t + h))$.

Now we consider the family of Cauchy problems

$$\left. \begin{array}{l} \partial_t u = a\Delta u - F(u, x, [\Lambda t + \omega]) - g(x, t) \\ \qquad = a\Delta u - F(u, x, \Lambda t + \omega) - g(x, t), \\ u|_{\partial\Omega} = 0, \quad u|_{t=\tau} = u_\tau, \quad u_\tau \in (L_2(\Omega))^N, \quad t \geq \tau, \\ G = (\omega, g(t)) \in T^k \times H(\phi). \end{array} \right\} \quad (A3.6)$$

It is clear that the problem (A3.6) is similar to the problem (A1.9), (A1.14). All the properties (A1.10) - (A1.13) and estimates (A1.15) - (A1.17) are valid for the problem (A3.6) with the same constants. So we define the family of processes $\{U_G(t, \tau)\}$, $G = (\omega, g) \in T^k \times H(\phi)$, acting on $(L_2(\Omega))^N$.

Lemma A3.2.
(1) The family $\{U_G(t, \tau)\}$, $G \in T^k \times H(\phi)$, is uniformly bounded on $(L_2(\Omega))^N$, i.e.

$$\|U_G(t, \tau)\| \leq R_0(\|u_\tau\|) \quad \forall t \geq \tau, \tau \in \mathbf{R}, G \in H(P).$$

(2) If $t > \tau$ then $U_G(t, \tau)u_\tau \in (H_1^0(\Omega))^N$ for any $G \in H(P)$, $u_\tau \in (L_2(\Omega))^N$. Moreover

$$\|U_G(t, \tau)u_\tau\|_1 \leq R_1(\|u_\tau\|, \delta) \quad \forall G \in H(P),$$

$$\forall t, \tau, \quad \tau + \delta \le t \le \tau + 1, \quad \tau \in \mathbf{R}, \quad \delta \in (0,1).$$

The proof is similar to the proof of Lemma A3.1.

Proposition A3.2.
The family $\{U_G(t,\tau)\}$, $G \in H(P)$ is uniformly compact and

$$((L_2(\Omega))^N \times H(P), (L_2(\Omega))^N)\text{-continuous.}$$

Proof   Due to (A1.15) the set

$$B_0 = \{u \in (L_2(\Omega))^N \mid \|u\|^2 \le \frac{2}{\gamma^2}(C_1 + \|\phi\|_{L_\infty}^2)\}$$

is uniformly absorbing. Lemma A3.2 implies that the set

$$B_1 = \{u \in (H_1^0(\Omega))^N \mid \|u\|_1^2 \le R_1^2\}, \quad \text{where}$$

$$R_1^2 = R(R_0, \delta), \quad R_0 = \frac{2}{\gamma^2}(C_1 + \|\phi\|_{L_\infty}^2),$$

is compact in $(L_2(\Omega))^N$ and uniformly absorbing. Hence, the family is uniformly compact.

The continuity property can be established using the same reasoning as for the Navier-Stokes system.

The family of processes $\{U_G(t,\tau)\}$, $G \in H(P)$, generates the semigroup $\{S(t)\}$ acting on $(L_2^2(\Omega))^N \times H(P)$:

$$S(t_1)(u_0, \omega, g) = (U_{(\omega,g)}(t_1, 0)u_0, [\Lambda t_1 + \omega], g^{t_1}).$$

Therefore Theorem A2.1 is applicable and we obtain the following result.

Theorem A3.2.
(1) The set $\mathcal{A} = \{(u(0), G) \mid u(t)$ is any bounded complete trajectory of the problem (A3.1) with arbitrary $G \in H(P)\}$ is the attractor of the semigroup $\{S(t)\}$ acting on $(L_2(\Omega))^N \times H(P)$.
(2) The set $\mathcal{A}_{H(P)} = \{u(0) \mid u(t)$ is any bounded complete trajectory of the problem (A3.6) with arbitrary symbol $G = (\omega, g(t)) \in T^k \times H(\phi)\}$ is

the uniform attractor of the family $\{U_G(t,\tau)\}$, $G \in H(P)$ corresponding to the problem (A3.6).

Up to the moment we've dealt with compact processes. Now we investigate an example with asymptotically compact processes, generated by a hyperbolic equation.

Example A3.3.   Time-dependent hyperbolic equation with almost periodic terms

We shall consider the equation (A1.20) where $\phi(x,t)$ is an almost periodic function with values in $L_2(\Omega)$ and $f(u,t)$, for simplicity, is a quasi-periodic function:

$$f(u,t) = F(u, \Lambda t) = F(u, \lambda_1 t, ..., \lambda_k t),$$

where

$$F(u, \omega_1, ..., \omega_j + 2\pi, ..., \omega_k) = F(u, \omega_1, ..., \omega_j, ..., \omega_k) \quad (j = 1, ..., k).$$

(The more general case will be sketched in §A6). We assume that the function $F(u,\omega)$ satisfies the conditions (A1.21) where $f(u,t)$ is replaced with $F(u,\omega)$, $\omega = \Lambda t$

$$\mathcal{F}(u,\omega) = \int_0^u F(v,\omega)\,dv, \quad u, \omega \in \mathbf{R} \times T^k,$$

and

$$\partial_t f = (F'_\omega, \Lambda), \quad \partial_t \mathcal{F} = (\mathcal{F}'_\omega, \Lambda).$$

We shall use the notation introduced in the second example. In particular, $H(\phi)$ is a hull of $\phi(x,t)$, $\phi(t) = (\omega_0(t), \phi(t))$, where $\omega_0(t) = [\Lambda t]$, $H(P) = T^k \times H(\Omega)$ (as in Example A3.2).

We consider the family of Cauchy problems

$$\partial_t^2 u + \gamma \partial_t u = \Delta u - F(u, [\Lambda t + \omega]) - g(x,t), \quad u|_{\partial\Omega} = 0,$$
$$u|_{t=\tau} = u_\tau, \quad \partial_t u|_{t=\tau} = p_\tau, \tag{A3.7}$$

$t \geq \tau$, $\tau \in \mathbf{R}$, $\omega \in T^k$, $g \in H(\phi)$. The problem (A3.7) has a unique solution and generates the family of processes $\{U_G(t,\tau)\}$, $G = (\omega, g) \in T^k \times H(\phi)$ acting on the space $E = H_1^0(\Omega) \times L_2(\Omega)$.

Lemma A3.3.

The family $\{U_G(t,\tau)\}$, $G \in T^k \times H(\phi)$ is uniformly bounded and the following inequality holds:

$$z(t) \leq z(\tau) \cdot \exp(-\beta(t - \tau)) + (C_0 + C_1 \|g\|_{L_\infty}), \qquad (A3.8)$$

where $L_\infty = L_\infty(\mathbf{R}, L_2(\Omega))$,

$$z(t) = \|u(t)\|_1^2 + \|v(t)\|^2 + C_2\langle \mathcal{F}(u,\omega), 1\rangle + C_3, \qquad (A3.9)$$

$v(t) = \partial_t u(t) + \alpha u(t)$, $\alpha > 0$, α is sufficiently small, C_3 is sufficiently large.

We recall that the proof of (A3.8) uses the substitution $v = \partial_t u + \alpha u$ in (A1.20). Taking the scalar product of the equation (A1.20) with $v(t)$ and using estimates based on (A1.21) we deduce (A3.8).

The right-hand side of (A3.8) doesn't depend on g, because $\|g\|_{L_\infty} = \|\phi\|_{L_\infty}$.

Proposition A3.3.

The family $\{U_G(t,\tau)\}$, $G \in H(P)$ is uniformly asymptotically compact and $(E \times H(P), E)$-continuous.

Proof   First we shall deduce that this family has a uniformly absorbing set, bounded in E.

Due to (A1.21) we have $|\mathcal{F}(u,\omega)| \leq C_4(1 + |u|^{\rho+2})$. Hence, according to the Sobolev embedding theorem,

$$|\langle \mathcal{F}(u,\omega), 1\rangle| \leq C_5|\langle 1 + |u|^{\rho+2}, 1\rangle| \leq C_6(\|u\|_1 + 1)^{\rho+2}. \qquad (A3.10)$$

(A3.9) and (A3.10) imply that $|z(\tau)| \leq C_7(R)$ if $\|y(\tau)\| \leq R$. Now if we take m in (A1.21) sufficiently small and C_3 in (A3.9) sufficiently large then we infer from (A3.9) that

$$z(t) \geq \frac{1}{2}\|u(t)\|_1^2 + C_8\|p(t)\|^2,$$

$$p(t) = \partial_t u(t), \quad y(t) = (u(t), \partial_t u(t)).$$

Therefore, according to (A3.8) the set

$$B_0 = \{(u,p) \mid \|u\|_1^2 + \|p\|^2 \leq 2(C_0 + C_1\|\phi\|_{L_\infty}^2)\}$$

is uniformly absorbing.

Now we check that the family $\{U_G(t,\tau)\}$, $G \in H(P)$ possesses a compact uniformly attracting set.

We can rewrite $U_G(t,\tau)y(\tau)$ in the form

$$U_G(t,\tau)y(\tau) = U_1(t,\tau)y(\tau) + U_2(t,\tau,G)y(\tau) + \tilde{y}(t,\tau,g). \qquad \text{(A3.11)}$$

Here U_1 is a linear process corresponding to the linear Cauchy problem

$$\left.\begin{array}{l} U_1(t,\tau)y(\tau) = (v(t),v_t(t)), \\ \partial_t^2 v + \gamma\partial_t v = \Delta v, \quad u|_{\partial\Omega} = 0, \quad v|_{t=\tau} = u(\tau), \quad \partial_t v|_{t=\tau} = p(\tau). \end{array}\right\} \tag{A3.12}$$

It's easy to show that any solution of (A3.12) decays exponentially

$$\|U_1(t,\tau)y(\tau)\| \le C\exp[-\delta(t-\tau)] \cdot \|y(\tau)\|, \qquad \text{(A3.13)}$$

where C doesn't depend on $\|y(\tau)\|$ or τ. In fact, $U_1(t,\tau) = U_1(t-\tau,0)$ is a semigroup.

The function $\tilde{g}(t,\tau,g) = (\tilde{u}(t), \partial_t\tilde{u}(t))$ is the solution of the inhomogeneous problem

$$\begin{array}{l} \partial_t^2\tilde{u}(t) + \gamma\partial_t\tilde{u}(t) = \Delta\tilde{u}(t) - g(t,x), \quad \tilde{u}|_{\partial\Omega} = 0, \\ \tilde{u}|_{t=\tau} = 0, \quad \partial_t\tilde{u}|_{t=\tau} = 0. \end{array} \tag{A3.14}$$

One can show that the solution of (A3.14) is given by Duhamel's integral:

$$\begin{aligned} \tilde{y}(t,\tau,g) &= \int_\tau^t U_1(t,\tau_1)\{0, g(\tau_1)\}\,\mathrm{d}\tau_1 \\ &= \int_0^{t-\tau} U_1(t-\tau-s,0)\{0, g^\tau(s)\}\,\mathrm{d}s. \end{aligned} \tag{A3.15}$$

For any $T > 0$ the set K_T defined by the formula

$$K_T = \left\{ \int_0^t U_1(t-s,0)\{0, g(s)\}\,\mathrm{d}s, t \in [0,T], g \in H(\phi) \right\}$$

is compact. Indeed, let

$$u_n = \int_0^{t_n} U_1(t_n - s,0)\{0, g_n(s)\}\,\mathrm{d}s$$

be any sequence in K_T, where $t_n \in [0,T]$, $g_n \in H(\phi)$. The set $[0,T] \times H(\phi)$ is compact in $\mathbf{R} \times C_b(\mathbf{R}, L_2(\Omega))$. Hence, there exists a sub-sequence $n_i \to$

$+\infty$ such that $g_{n_i}(s) \to g(s)$ $(n_i \to +\infty)$ uniformly with respect to $s \in \mathbf{R}$ and $t_{n_i} \to t$ $(n_i \to +\infty)$. Therefore, according to (A3.13), we get

$$u_{n_i} = \int_0^{t_i} U_1(t_i - s, 0)\{0, g_{n_i}(s)\} \, ds$$

$$\to \int_0^t U_1(t - s, 0)\{0, g(s)\} \, ds = u,$$

where $u \in K_T$, i.e. K_T is compact. Let's show that the set $K_1 = \bigcup_{T>0} K_T$ is precompact too. It is clear that (A3.13) implies

$$\| \int_0^{T_1} U_1(T_1 - s, 0)\{0, g(s)\} \, ds - \int_0^T U_1(T - s, 0)\{0, g^{T_1 - T}(s)\} \, ds \|$$

$$= \| \int_0^{T_1 - T} U_1(T_1 - s, 0)\{0, g(s)\} \, ds + \int_{T_1 - T}^{T_1} U_1(T_1 - s, 0)\{0, g(s)\} \, ds$$

$$- \int_0^T U_1(T - s, 0)\{0, g^{T_1 - T}(s)\} \, ds \|$$

$$= \| \int_0^{T_1 - T} U_1(T_1 - s, 0)\{0, g(s)\} \, ds$$

$$+ \int_0^T U_1(T - s_1, 0)\{0, g(T_1 - T + s_1)\} \, ds_1$$

$$- \int_0^T U_1(T - s, 0)\{0, g^{T_1 - T}(s)\} \, ds \|$$

$$= \| \int_0^{T_1 - T} U_1(T_1 - s, 0)\{0, g(s)\} \, ds$$

$$+ \int_0^T U_1(T - s_1, 0)\{0, g^{T_1 - T}(s_1)\} \, ds_1$$

$$- \int_0^T U_1(T - s, 0)\{0, g^{T_1 - T}(s)\} \, ds \|$$

$$= \| \int_0^{T_1 - T} U_1(T_1 - s, 0)\{0, g(s)\} \, ds \|$$

$$\le C_1 \exp(-\delta T) \quad \forall T_1 \ge T.$$

Here we have replaced s with $s = T_1 - T + s_1$ and used inequality (A3.13). Hence

$$\mathrm{dist}(K_1, K_T) \le C_1 \exp(-\delta T) \to 0 \quad (T \to +\infty).$$

Thus the set K_1 can be approximated by compact sets, so K_1 is precompact.

We conclude from (A3.15) that

$$\tilde{y}(t,\tau,g) \in K_1 \quad \text{for any } g \in H(\phi), t \geq \tau, \tau \in \mathbf{R}.$$

Finally the function

$$U_2(t,\tau,G)y(\tau) = (w(t), \partial_t w(t))$$

satisfies the inhomogeneous equation

$$\begin{aligned}
\partial_t{}^2 w + \gamma \partial_t w &= \Delta w - F(u(t), \Lambda t + \omega), \\
w|_{t=\tau} &= 0, \quad \partial_t w|_{t=\tau} = 0.
\end{aligned} \tag{A3.16}$$

We have seen above $u(t) \in C_b(\mathbf{R}_\tau, H_1)$, so due to (A1.21) $F(u(t), \Lambda t + \omega) \in C_b(\mathbf{R}_\tau, H)$; moreover, the norm of this function is uniformly bounded when $\|(u(\tau), \partial_t u(\tau))\|$ is bounded. (See (A3.8)). Hence

$$(w(t), \partial_t w(t)) \in C_b(\mathbf{R}_\tau, E)$$

and the norm of this function is uniformly bounded in $C_b(\mathbf{R}_\tau, E)$. Then by differentiation of (A3.16) we obtain that $p(t) = \partial_t w(t)$ is a solution of the equation

$$\begin{aligned}
\partial_t{}^2 p + \gamma \partial_t p - \Delta p &= -F'_\omega(u, \Lambda t + \omega)\Lambda - F'_u(u, \Lambda t + \omega)u'(t), \\
p|_{t=\tau} &= 0, \quad \partial_t p|_{t=\tau} = -F(u(\tau), \Lambda \tau + \omega).
\end{aligned} \tag{A3.17}$$

Since $u \in C_b(\mathbf{R}_\tau, H_1)$ and $|\partial_\omega F| \leq C(1 + |u|)^{\rho+3}$, we infer that

$$F'_\omega(u(t), \Lambda t + \omega)\Lambda \in C_b(\mathbf{R}_\tau, H_{-1+\sigma}), \quad \sigma = 1 - \frac{\rho}{2}.$$

Then, since $|\partial_u F| \leq C(1 + |u|)^\rho$, the function $\partial_u F(u, \Lambda t + \omega)u'(t)$ belongs to $C_b(\mathbf{R}_\tau, H_{-1+\sigma})$, where $\sigma = 1 - \rho/2$ (see Temam [2], Babin & Vishik [1]). Thus the right-hand side of (A3.17) belongs to $C_b(\mathbf{R}_\tau, H_{-1+\sigma})$ and its norm in this space is uniformly bounded with respect to $g \in H(\phi)$, $\omega \in T^k$, $\|y(\tau)\| \leq R$.

We infer from (A3.8) applied to $q = A^{(\sigma-1)/2}p$ (where $Au = \Delta u$, $u|_{\partial\Omega} = 0$), that

$$(p, p') = (w', w'') \in C_b(\mathbf{R}_\tau, H_\sigma \times H_{\sigma-1}). \tag{A3.18}$$

We then return to (A3.16); since $F(u, \Lambda t + \omega) \in C_b(\mathbf{R}_\tau, H)$ we find using (A3.18), that

$$\Delta w \in C_b(\mathbf{R}_\tau, H_{\sigma-1}), \quad \text{i.e. } w \in C_b(\mathbf{R}_\tau, H_{1+\sigma}).$$

So we have proved that $(w(t), \partial_t w(t)) \in C_b(\mathbf{R}_\tau, H_{\sigma+1} \times H_\sigma)$ and the set $\{(w(t), \partial_t w(t)) \mid \|y(\tau)\| \leq R, t \geq \tau\}$ is uniformly bounded in $H_{\sigma+1} \times H_\sigma$ with respect to $\omega \in T^k$, $g \in H(\phi)$. In other words the set

$$K_2 = \bigcup_{\tau \in \mathbf{R}} \bigcup_{t \geq \tau} \{U_2(t, \tau, G)B_0 \mid G \in H(P)\}$$

is precompact in E, where B_0 is the absorbing set of the family $\{U_g(t, \tau)\}$, $G \in H(P)$.

Finally we'll prove that the compact set $K = \overline{K_1 + K_2}$ is uniformly attracting with respect to the family $\{U_G(t, \tau)\}$, $G \in H(P)$. Note that for an arbitrary bounded set $B \in \mathcal{B}(E)$ there exists a moment $T = T(B)$ such that $U_G(t, \tau)B \subseteq B_0$ when $t - \tau \geq T$, $G \in H(P)$. Thus it is sufficient to consider only the case $B = B_0$.

So let $y(\tau) \in B_0$; then due to (A3.11)

$$
\begin{aligned}
\text{dist}(U_G(t, \tau)y(\tau), & K_1 + K_2) \\
& \leq \text{dist}(U_G(t, \tau)y(\tau), U_2(t, \tau, G)y(\tau) + \tilde{y}(t, \tau, g)) \\
& = \|U_1(t, \tau)y(\tau)\| \\
& \leq C \exp(-\delta(t - \tau)) \cdot \|y(\tau)\| \to 0 \quad (t - \tau \to +\infty).
\end{aligned}
$$

Hence $K = \overline{K_1 + K_2}$ attracts B_0.

Using standard techniques one can prove that the family of processes $\{U_G(t, \tau)\}$, $G \in H(P)$ is $(E \times H(P), E)$-continuous.

Now if we introduce the semigroup $\{S(t)\}$, corresponding to the family of processes $\{U_G(t, \tau)\}$, $G \in H(P)$, and acting on $E \times H(P)$ then we get the following theorem.

Theorem A3.3.

(1) The set $\mathcal{A} = \{(y(0), G) \mid y(t)$ is any complete bounded trajectory of the problem (A3.7) with arbitrary $G \in H(P)\}$ is the attractor of the semigroup $\{S(t)\}$ acting on $E \times H(P)$.

(2) The set $\mathcal{A}_{H(P)} = \{u(0) \mid u(t)$ is any complete bounded trajectory of the problem (A3.7) with arbitrary symbol $G \in H(P)\}$ is the uniform attractor of the family of processes $\{U_G(t, \tau)\}$, $G \in H(P)$.

§A4   Upper bounds for the dimension of attractors of non-autonomous dynamic systems with quasi-periodic terms

As we have seen in §A2 the uniform attractor $\mathcal{A}_1$ of the family of processes $\{U_G(t, \tau)\}$, $G \in H(P)$, coincides with the projection $\pi_1(\mathcal{A})$ of the attractor $\mathcal{A}$ of the corresponding semigroup $\{S(t)\}$ acting on $E \times H(P)$. Hence, for Hausdorff dimension we have dim $\mathcal{A}_1 \leq$ dim $\mathcal{A}$.

First we recall the formulas for upper bound for the Hausdorff dimension of $\mathcal{A}$ and then we apply these estimates for $\dim_H \mathcal{A}_1$ in the case of a quasi-periodic process.

Let $\{S(t) \mid t \geq 0\}$ be a semigroup acting on a Hilbert space H. Let $X \subseteq H$ be a compact set. A (nonlinear) continuous mapping S from X into H is called uniformly quasi-differentiable on X if for every $u \in X$ there exists a linear operator $S'(u) \in \mathcal{L}(H, H)$ (quasi-differential) such that for any $u_1 \in X$ the inequality

$$||S(u_1) - S(u) - S'(u)(u_1 - u)||_H \leq \gamma(||u_1 - u||_H) \cdot ||u_1 - u||_H \quad (A4.1)$$

holds, where $\gamma(\xi) \to 0$ $(\xi \to 0^+)$, γ doesn't depend on $u_1, u \in X$.

We assume that for $u_0 \in X$ the quasi-derivative $S'(t, u_0)$ of a map $S(t)$ on X is generated by the variational equation

$$\partial_t v + L(u(t))v = 0, \quad v|_{t=0} = \xi, \quad (A4.2)$$

where $u(t) = S(t)u_0$, i.e. $S'(t, u_0)\xi = v(t)$.

It is assumed also that the operator $L(u(t))$ maps a dense subset $H_2 \subset H$ into H, $L(u(t)) : H_2 \to H$, and the equation (A4.2) generates the family of continuous maps $S'(t, u_0) : H \to H$, where $S'(t, u_0)$ is the quasi-derivative of S in $u_0 \in X$.

Let $d \in \mathbb{N}$ and $L : H_2 \to H$. The number

$$Tr_d L = \inf_Q Tr LQ \quad (A4.3)$$

is called the d-trace of the linear operator L, where the infimum is taken over all orthogonal projections Q of H onto QH having dimension d. We recall that

$$TrLQ = \sum_{j=1}^{d} (L\phi_j, \phi_j)_H, \quad (A4.4)$$

where $\phi_1, ..., \phi_d \in QH \subseteq H_2$ are an orthogonal basis in the space QH.

We are going to use the following theorem by Constantin, Foias & Temam [1] for estimation of the Hausdorff dimension of attractors.

Theorem A4.1.
Let $S(t)X = X$ $\forall t \geq 0$, $S(t)$ be uniformly quasi-differentiable on X and $\|S'_u(t;u)\|_{\mathcal{L}(H,H)} \leq C(t)$ for every $t \geq 0$. Let q_d be

$$q_d = \lim_{t \to +\infty} \sup_{u_0 \in X} \left\{ -\frac{1}{t} \int_0^t Tr_d L(u(s))\, ds \right\}, \qquad (A4.5)$$

where $u(t) = S(t)u_0$. Then the following estimate holds:

$$\dim_H X \leq d, \ \text{if } q_d < 0. \qquad (A4.6)$$

For estimating $Tr_d(L(u(t)))$ in (A4.5) the following comparison principle is convenient. (See Ilyashenko [1], [2], Babin & Vishik [1], [5]).

Lemma A4.1.
Let L_0 be a self-adjoint positive operator on H with dense domain H_2 having eigenvalues $0 < \lambda_1 \leq \lambda_2 \leq \ldots$. It is assumed that

$$\begin{aligned} \langle L(u(t))v, v \rangle \geq &\langle L_1(t)v, v \rangle \\ &\equiv \nu_0 \langle L_0 v, v \rangle - h(t,u_0)\langle v, v \rangle, \end{aligned} \qquad (A4.7)$$

where $L_1(t) = \nu_0 L_0 - h(t, u_0)I$. Then

$$\begin{aligned} \liminf_{t \to +\infty} &\sup_{u_0 \in X} \left\{ -\frac{1}{t} \int_0^t Tr_d L(u(s))\, ds \right\} \\ &\leq \lim_{t \to +\infty} \sup_{u_0 \in X} \left\{ -\frac{1}{t} \int_0^t Tr_d(L_1(s))\, ds \right\} \\ &= -\nu_0 \sum_{j=1}^{d} \lambda_j + d \cdot \limsup_{t \to +\infty} \sup_{u_0 \in X} \left\{ -\frac{1}{t} \int_0^t h(s, u_0)\, ds \right\}. \end{aligned} \qquad (A4.8)$$

The proof of (A4.8) is given in Babin & Vishik [1].

Let A be a linear operator, $A: H_2 \to H$, where $\{H_s\}$ is the scale of spaces corresponding to the positive self-adjoint operator L_0, $L_0: H_2 \to H$.

We assume that we are given a nonlinear operator $\mathcal{F}(u,\omega)$, where $u \in H_2$, $\omega \in T^k$, which maps $H_2 \times T^k$ into H.

Now we consider a quasi-periodic non-autonomous Cauchy problem:

$$\partial_t u = -Au - \mathcal{F}(u,[\Lambda t + \omega_0]),$$
$$u|_{t=\tau} = u_\tau, \quad \Lambda = (\lambda_1, ..., \lambda_k), \omega_0 \in T^k. \tag{A4.9}$$

We assume that the problem (A4.9) generates the family of processes $\{U_{\omega_0}(t,\tau)\}$, $\omega_0 \in T^k$, acting on H. As usual we define a semigroup $\{S(t)\}$, acting on $H \times T^k$,

$$S(t)(u_0,\omega_0) = (U_{\omega_0}(t,0)u_0,[\Lambda t + \omega_0]).$$

It is clear that the semigroup $\{S(t)\}$ can be constructed by use of the following autonomous dynamic system:

$$\partial_t u = -Au - F(u,\omega),$$
$$\partial_t \omega = \Lambda, \tag{A4.10}$$
$$u|_{t=0} = u_0, \quad \omega|_{t=0} = \omega_0.$$

Or in a short form
$$\partial_t y = -\mathcal{B}(y), \quad y|_{t=0} = y_0, \tag{A4.10'}$$

where $y = (u,\omega)$, $\mathcal{B}(y) = (Au + F(u,\omega),-\Lambda)$.

We assume that the family of processes $\{U_{\omega_0}(t,\tau)\}$, $\omega_0 \in T^k$, is uniformly asymptotically compact and $(H \times T^k, H)$-continuous. Due to Theorem A2.1 the semigroup $\{S(t)\}$ possesses an attractor $\mathcal{A}$ compact in $E = H \times T^k$.

We make the following assumptions concerning the semigroup $\{S(t)\}$:

(1) the semigroup $\{S(t)\}$ is uniformly quasi-differentiable on $\mathcal{A}$ (see (A4.1));

(2) the quasi-derivative $S'(t,y_0)z_0 = z(t)$ satisfies the variational equation

$$\partial_t z(t) = -\mathcal{B}'(y(t))z(t), \quad z|_{t=0} = z_0. \tag{A4.11}$$

Or in complete form for $z(t) = (v(t),\mu)$:

$$\partial_t v(t) = -Av(t) - F'_u(u(t),\omega(t))v(t) - F'_\omega(u(t),\omega(t))\mu,$$
$$\partial_t \mu = 0, \tag{A4.12}$$

where $u(t) = U_{\omega_0}(t,\tau)u_0$, $\omega(t) = [\Lambda t + \omega_0]$, $\mu \in \mathbf{R}^k$.

(3) The system (A4.11) (or (A4.12)) satisfies the hypothesis of Lemma A4.1, i.e. the following estimates hold:

$$(B'(y(t))z(t), z(t))$$
$$= (L_0 v, v) + \langle F'_u(u(t), \omega(t))v, v\rangle + \langle F'_\omega(u(t), \omega(t))\mu, v\rangle$$
$$\geq \langle L_0 v, v\rangle - h(t)\langle v, v\rangle - h_1(t)\mu \cdot \mu$$
$$\equiv \langle L_1(t)(v, \mu), (v, \mu)\rangle, \tag{A4.13}$$

$$L_1(t) \equiv (L_0 - h(t)I, -h_1(t)I_k), \quad L_0 = A, \nu_0 = 1.$$

To reduce the estimate (A4.13) to the form (A4.7) one has to replace v in (A4.7) by the pair $(v, \mu) \in H_2 \times \mathbf{R}^k$ and the operator $L_1(t)$ by the operator $(L_0 - h(t)I, -h_1(t)I_k) : H_2 \times \mathbf{R}^k \to H \times \mathbf{R}^k$. Let us show that after these substitutions the estimate (A4.8) implies that for large enough d (see (A4.15))

$$q_d = \lim_{t\to+\infty} \sup_{y_0 \in \mathcal{A}} \left\{ -\frac{1}{t}\int_0^t Tr_d \mathcal{B}'(y(t_1))\, dt_1 \right\}$$

$$\leq -(\lambda_1 + ... + \lambda_{d-k}) + (d-k)\limsup_{t\to+\infty}\left\{\frac{1}{t}\int_0^t h(t_1)\, dt_1\right\} \tag{A4.14}$$

$$+ k\limsup_{t\to+\infty}\left\{\frac{1}{t}\int_0^t h_1(t_1)\, dt_1\right\}\lambda_i \to +\infty \quad \text{as} \quad i \to \infty,$$

where $\lambda_1, ..., \lambda_{d-k}$ are the first $d-k$ eigenvalues of the operator L_0. Indeed, the self-adjoint operator $L_1(t) = (L_0 - h(t)I, -h_1(t)I_k)$, acting on $H \times \mathbf{R}^k$ (with dense domain $H_2 \times \mathbf{R}^k$ in $H \times \mathbf{R}^k$) has the following eigenvalues:

$$\lambda_i - h(t) \quad (i = 1, 2, ...) \quad \text{and} \quad -h_1(t) \text{ having}$$
$$\text{the multiplicity } k.$$

So, if d is large enough that

$$\lambda_{d-k} - h(t) > -h_1(t) \quad \forall t \geq 0, \tag{A4.15}$$

then due to the Courant principle the infimum

$$\inf_Q TrL_1(t)Q = Tr_d L_1(t),$$

where Q is an orthogonal projection in $H \times \mathbf{R}^k$ onto a finite-dimensional space in $H_2 \times \mathbf{R}^k$ having dimension d, is attained on the space spanned by the first d eigenvectors of the operator $L_1(t)$. According to (A4.15), the corresponding eigenvalues are

$$\lambda_1 - h(t), ..., -h_1(t), ..., -h_1(t) \; k \text{ times}, ..., \lambda_{d-k} - h(t). \qquad \text{(A4.16)}$$

Hence for $d - k > 0$,

$$\liminf_{t \to +\infty} \left\{ -\frac{1}{t} \int_0^t T_{r_d} L_1(t_1) \, dt_1 \right\} = -(\lambda_1 + ... + \lambda_{d-k})$$

$$+ (d-k) \limsup_{t \to +\infty} \left\{ \frac{1}{t} \int_0^t h(t_1) \, dt_1 \right\}$$

$$+ k \limsup_{t \to +\infty} \left\{ \frac{1}{t} \int_0^t h_1(t_1) \, dt_1 \right\}. \qquad \text{(A4.17)}$$

Now (A4.14) follows from Lemma A4.1.

Finally, to get $\dim_H \mathcal{A} \le d$ (and respectively $\dim_H \mathcal{A}_1 \le d$) we need $q_d < 0$. Thus, due to Lemma A4.1 and (A4.17) it is sufficient to have

$$-(\lambda_1 + ... + \lambda_{d-k})$$

$$+ (d-k) \limsup_{t \to +\infty} \left\{ \frac{1}{t} \int_0^t h(t_1) \, dt_1 \right\} + k \limsup_{t \to +\infty} \left\{ \frac{1}{t} \int_0^t h_1(t_1) \, dt_1 \right\} < 0. \qquad \text{(A4.18)}$$

§A5 On finite-dimensionality of attractors on non-autonomous equations in mathematical physics with quasi-periodic terms

1. Navier-Stokes equations ($n = 2$) with quasi-periodic exterior forces.

We consider a two-dimensional Navier-Stokes system (A3.1) with quasi-periodic exterior force

$$g(t, x) = G(x, \omega(t)), \quad \omega(t) = [\Lambda t + \omega_0], \quad G(x, \omega) \in C(\Omega \times T^k), \quad \text{(A5.1)}$$

where $\Lambda = (\lambda_1, ..., \lambda_k)$, $\omega_0 = (\omega_{01}, ..., \omega_{0k}) \in T^k$,

$$G(x, \omega_1, ..., \omega_j + 2\pi, ..., \omega_k) = G(x, \omega_1, ..., \omega_j, ..., \omega_k)$$

is a 2π-periodic function with respect to each ω_j ($j = 1, ..., k$). For simplicity we assume that $\nu = 1$. In this case the system (A4.10) has the form

$$\partial_t u = -Lu - B(u, u) + G(x, \omega), \quad \partial_t \omega = \Lambda,$$
$$u|_{t=0} = u_0, \quad \omega|_{t=0} = \omega_0. \qquad \text{(A5.2)}$$

Or briefly in the form (4.10'), where

$$-\mathcal{B}(y) = (-Lu - B(u,u) + G(x,\omega), \Lambda), \quad y = (u,\omega).$$

Let us consider the variational equations of the system (A5.2):

$$\left. \begin{aligned} \partial_t v &= -Lv - B(u(t),v) - B(v,u(t)) + G'_\omega(x,\omega(t))\mu, \\ \partial_t \mu &= 0, \quad \mu = (\mu_1,...,\mu_k) \in \mathbf{R}^k, \quad v = v(x,t) = v(t), \end{aligned} \right\} \tag{A5.3}$$

or briefly

$$\partial_t z = -\mathcal{B}'(y(t))z, \quad z = (v,\mu). \tag{A5.4}$$

Here $u(t) = U_{\omega_0}(t,0)u_0$ is the solution of (A5.2) with initial data $u(0) = u_0$, $\omega(0) = \omega_0$. We can also use (A5.4), where

$$-\mathcal{B}'(y(t))z = (-Lv - B(u(t),v) - B(v,u(t)) + G'_\omega(x,\omega(t))\mu, 0).$$

Let's write the estimate (A4.13) for this case:

$$\begin{aligned} \langle \mathcal{B}'(y(t)z,z) \rangle =& (Lv,v) + (B(u(t),v),v) \\ &+ (B(v,u(t)),v) + \langle G'_\omega(x,\omega(t))\mu, v \rangle \\ \geq& \frac{1}{2}\|v\|_1^2 - \left(\frac{b}{2}\gamma(t) + C_1\|u(t)\|_1^2 \right)\|v\|^2 - \frac{1}{2b}\gamma(t)\mu^2. \end{aligned} \tag{A5.5}$$

In the inequality above we have introduced the notation

$$\gamma(t) = \gamma(t,\omega_0) = \|G'_\omega(\cdot,\omega(t))\|,$$

and we have used the equality $(B(u(t),v),v) = 0$ and the well-known estimate

$$|(B(v,u),v)| \leq C\|v\|_1\|v\| \cdot \|u\|_1. \tag{A5.6}$$

The inequality (A5.5) implies (A4.13) with

$$\begin{aligned} L_0 &= -\frac{1}{2}\Pi\Delta, \quad h(t) = h(t,u_0,\omega_0) = C_1\|u(t)\|_1^2 + \frac{b}{2}\gamma(t), \\ h_1(t) &= h_1(t,\omega_0) = \frac{1}{2b}\gamma(t). \end{aligned} \tag{A5.7}$$

Due to (A4.13), (A4.14) we get

$$\begin{aligned} q_d \leq& -\frac{1}{2}(\lambda_1 + ... + \lambda_{d-k}) \\ &+ (d-k)\limsup_{t \to +\infty} \sup_{\omega_0,u_0} \left\{ \frac{1}{t}\int_0^t C_1\|u(t)\|_1^2 \, dt + \frac{b}{2} \cdot \frac{1}{t}\int_0^t \gamma(t)\,dt \right\} \\ &+ k \cdot \frac{1}{2b}\limsup_{t \to +\infty} \sup_{\omega_0} \left\{ \frac{1}{t}\int_0^t \gamma(t)\,dt \right\} \quad (\omega_0 \in T^k, u_0 \in \mathcal{A}_1). \end{aligned} \tag{A5.8}$$

According to the Kronecker-Weyl theorem for quasi-periodic functions

$$
\begin{aligned}
\limsup_{t\to+\infty} \sup_{\omega_0} \left\{ \frac{1}{t} \int_0^t \gamma(t)\,dt \right\} &= \lim_{t\to+\infty} \left\{ \frac{1}{t} \int_0^t \|G'_\omega(\cdot,\omega(t))\|\,dt \right\} \\
&= \frac{1}{|T^k|} \int_{T^k} \|G'_\omega(\cdot,\omega)\|\,d\omega = \sigma'.
\end{aligned}
\tag{A5.9}
$$

Using an analogue of (A1.8) and the Kronecker-Weyl theorem we have

$$
\begin{aligned}
\limsup_{t\to+\infty} \sup_{\omega_0,u_0} &\, C_1 \left\{ \frac{1}{t} \int_0^t \|u\|_1\,dt \right\} \\
&\le \lim_{t\to+\infty} \sup_{\omega_0,u_0} C_2 \left\{ \frac{1}{t} \int_0^t \|G(\cdot,\omega(\tau)\|_{-1}^2\,d\tau + \frac{1}{t}\|u(0)\|^2 \right\} \\
&= \frac{1}{|T^k|} \int_{T^k} \|G(\cdot,\omega)\|_{-1}^2\,d\omega \equiv C_2\sigma.
\end{aligned}
\tag{A5.10}
$$

We then apply the Cattabriga theorem:

$$
-\frac{1}{2}(\lambda_1 + \ldots + \lambda_{d-k}) \le -a^2(d-k)^2,
\tag{A5.11}
$$

where $a^2 > 0$. Taking into account (A5.9), (A5.10), (A5.11) we obtain from (A5.8)

$$
q_d \le -a^2(d-k)^2 + (\frac{b}{2}\sigma' + C_2\sigma)(d-k) + \frac{1}{2b}\sigma' k.
\tag{A5.12}
$$

The value $k^{1/2}/(d-k)^{1/2} = b$ minimizes the right-hand side of (A5.12). To estimate d for which $q_d < 0$ we consider the equality instead of the inequality in (A5.12); moreover b is considered to have the optimal value

$$
-a^2(d-k)^2 + \sigma' \cdot k^{1/2}(d-k)^{1/2} + C_2\sigma(d-k) = 0.
$$

It is easy to show that if k is large enough then

$$
d = k + \left(\frac{\sigma'}{a^2}\right)^{2/3} k^{1/3} + \mathcal{O}(1).
\tag{A5.13}
$$

Hence, due to (A4.6), (A4.14), (A4.17), (A4.18), (A5.8), (A5.12) and (A5.13) one has

$$
\dim_H \mathcal{A}_1 \le k + \left(\frac{\sigma'}{a^2}\right)^{2/3} k^{1/3} + \mathcal{O}(1).
\tag{A5.14}
$$

This estimate implies that when the number k of rationally independent frequencies $\lambda_1, ..., \lambda_k$ is growing, and σ and σ' are bounded, then the right-hand side of (A5.14) (estimating the dimension of $\mathcal{A}_1$) is growing too.

At the same time it is evident that

$$\dim_H \mathcal{A} \geq \dim_H \pi_2(\mathcal{A}) = \dim_H T^k = k.$$

One can construct a great variety of examples of exterior forces $G(t, x)$, for which $\dim_H(\mathcal{A}_1) \geq k$. This shows that the estimate (A5.1) is exact with respect to large k.

2. Time-dependent reaction-diffusion system with quasi-periodic terms.

We consider a particular case of the system (A3.6):

$$\partial_t u = a\Delta u - F(u, x, \omega(t)) - G(x, \omega(t)), \quad u\,|_{\partial\Omega} = 0,$$
$$u\,|_{t=\tau} = u_\tau, \quad u_\tau \in (L_2(\Omega))^N, \quad \omega_0 \in T^k. \qquad (A5.15)$$

Here $\omega(t) = [\Lambda t + \omega_0]$, $\omega_0 = (\omega_{01}, ..., \omega_{0k}) \in T^k$, vector functions $F(u, x, \omega)$ and $G(x, \omega)$ are periodic with period 2π with respect to each ω_j ($j = 1, ..., k$). The function $F(u, x, \omega)$ satisfies the conditions (A3.5), (A1.10) - (A1.13) and $G(x, \omega) \in C(\Omega \times T^k)$, $G'_{\omega_i} \in C(\Omega \times T^k)$ ($i = 1, ..., k$); F'_ω satisfies (A1.12). For simplicity let $p = 2n/(n-2)$, $n \geq 3$.

As we know from §A3 the system (A5.15) generates the family of processes $\{U_{\omega_0}(t, \tau)\}$, $\omega_0 \in T^k$, acting on $(L_2(\Omega))^N$. Due to Theorem A3.2 this family has the compact uniform attractor $\mathcal{A}_{T^k} = \mathcal{A}_1$. The problem is to estimate $\dim_H \mathcal{A}_1$.

We can write the autonomous system (A4.10) corresponding to (A5.15) in the form

$$\left.\begin{array}{l} \partial_t u = a\Delta u - F(u, x, \omega) - G(x, \omega), \quad u\,|_{\partial\Omega} = 0, \\ \partial_t \omega = \Lambda, \quad u\,|_{t=0} = u_0, \quad \omega\,|_{t=0} = \omega_0, \end{array}\right\} \qquad (A5.16)$$

or for brevity

$$\partial_t y = -My, \quad y\,|_{t=0} = y_0, \quad y = (u, \omega),$$
$$My = (-a\Delta u + F(u, x, \omega) + G(x, \omega), \Lambda). \qquad (A5.17)$$

The variational equations (A4.11) for the system (A5.16) are the following:

$$\left.\begin{array}{l} \partial_t v = a\Delta v - F'_u(u(t), x, \omega(t))v - F'_\omega(u(t), x, \omega(t))\mu \\ \qquad - G'_\omega(x, \omega(t))\mu, \\ \partial_t \mu = 0, \quad v\,|_{\partial\Omega} = 0, v\,|_{t=0} = v_0, \quad \mu\,|_{t=0} = \mu_0. \end{array}\right\} \qquad (A5.18)$$

We omit the proof of the differentiability of operators $S(t)(y_0)$ ($y_0 = (u_0, \omega_0)$) of the semigroup, corresponding to (A5.16), and that

$$S'(t, y_0)z_0 = (v(t), \mu_0) = z(t),$$

where $(v(t), \mu_0)$ is the solution of (A5.18). This is a well-known technical point. (See Babin & Vishik [1], Temam [2].)

We'll obtain the estimate (A4.13) for this case:

$$\langle M'_y(y(t))z, z \rangle = \langle a\Delta v, \nabla v \rangle + \langle F'_u(\cdot, \cdot, \cdot)v, v \rangle \\ + \langle F'_\omega(\cdot, \cdot, \cdot)\mu, v \rangle + \langle G'_\omega(\cdot, \cdot)\mu, v \rangle. \tag{A5.19}$$

The conditions (A1.11) - (A1.13) imply

$$\langle F'_u(\ )v, v \rangle \geq -C\langle v, v \rangle,$$
$$\langle F'_\omega(\ )\mu, v \rangle \geq -C\langle (1 + |u|^{(n+2)/(n-2)}), |v| \rangle |\mu| \\ \geq -C_2 \|1 + |u|^{(n+2)/(n-2)}\|_{L_{2n/(n+2)}} \cdot \|v\|_1 \cdot |\mu| \\ \geq -C_3(1 + \|u\|_{L_{2n/(n-2)}}^{(n+2)/(n-2)})\|v\|_1 \cdot |\mu| \\ \geq -\frac{1}{2}\langle a\nabla v, \nabla v \rangle - C_4(1 + \|u\|_1^{2(n+2)/(n-2)})\mu^2. \tag{A5.21}$$

Since $G'_{\omega_i}(x, \omega) \in C(\Omega \times T^k)$,

$$\langle G'_\omega(x, \omega)\mu, v \rangle \geq -\|G'_\omega(\cdot, \omega(t))\| \cdot |\mu| \cdot \|v\| \\ \geq -\frac{b}{2}\gamma_1(t)\|v\|^2 - \frac{1}{2b}\gamma_1(t)\mu^2, \tag{A5.22}$$

where we use notation $\gamma_1(t) = \|G'_\omega(\cdot, \omega(t))\|$.

We infer from (A5.18) - (A5.22) the following estimate:

$$-\langle M'_y(y)z, z \rangle \leq -\frac{\alpha}{2}\|v\|_1^2 + \left(C + \frac{b}{2}\gamma_1(t)\right)\|v\|^2 \\ + \left[C_4(\|u(t)\|_1^{2(n+2)/(n-2)} + 1) + \frac{1}{2b}\gamma_1(t)\right]\mu^2. \tag{A5.23}$$

We have used the fact that $(a\nabla v, \nabla v) \geq \alpha\|v\|_1$. Due to (A1.15) - (A1.17) if $u(t) \in \mathcal{A}_1$ one can prove that

$$\|u(t)\|_1^{2(n+2)/(n-2)} \leq C_5 \mathcal{G} + C_6, \\ \mathcal{G} = \sup_\omega \|G(\cdot, \omega)\|^{2(n+2)/(n-2)}). \tag{A5.24}$$

Similarly to (A5.8) - (A5.12) the inequalities (A5.23), (A5.24) imply

$$q_d \le -\frac{\alpha}{2}(\lambda_1 + \dots + \lambda_{d-k}) + \limsup_{t\to+\infty}\left\{\frac{1}{t}\int_0^t \left(C_1 + \frac{b}{2}\gamma_1(\tau)\right)d\tau\right\}(d-k)$$
$$+ \limsup_{t\to+\infty}\left\{\frac{1}{t}\int_0^t \left(C_4\,[(C_5\mathcal{G} + C_6) + 1] + \frac{1}{2b}\gamma_1(\tau)\right)d\tau\right\}k,$$

(A5.25)

where λ_i are the smallest eigenvalues of the operator $-\Delta u$ with $u\,|_{\partial\Omega}= 0$. Using the asymptotic behaviour $\lambda_i \ge C_0 i^{2/n}$ and formulae like (A5.9) one can get

$$q_d \le -\beta(d-k)^{1+2/n} + (d-k)\left(C + \frac{b}{2}\sigma'\right) + k\left(C_7\sigma + C_8 + \frac{1}{2b}\sigma'\right),$$

(A5.26)

where

$$\beta = \frac{\alpha}{2}C_0, \quad C_7 = C_4\cdot C_5, \quad C_8 = C_4\cdot C_6 + 1,$$

$$\sigma' = \frac{1}{|T^k|}\int_{T^k} \|G_\omega(\cdot,\omega)\|\,d\omega.$$

Taking the optimal value $b = (k(d-k))^{1/2}$ we can find the smallest $d > k$ such that $q_d < 0$. This d is the root of the equation, following from (A5.26),

$$-\beta(d-k)^{1+2/n} + C(d-k) + \sigma'(d-k)^{1/2}k^{1/2} + C_9 k = 0,$$

where $C_9 = C_7\sigma + C_8$. Solving the equation one has

$$d = k + C_{10}k^{n/(n+2)} + \dots \tag{A5.27}$$

Hence, according to (A4.6) we infer

$$\dim_H \mathcal{A}_1 \le k + C_{10}k^{n/(n+2)} + \dots \tag{A5.28}$$

Therefore, the upper bound for the Hausdorff dimension of the attractor $\mathcal{A}_1$ is growing when k is growing.

§A6   Sections of attractors and their dimensionality. The case of almost periodic terms

In §A5 we have proved that the dimension of the attractor for the quasi-periodic dynamic systems is of order k for large k, where k is the number of rationally independent frequencies of the corresponding quasi-periodic symbol. Thus the dimension of the attractor tends to infinity when k does. A general almost periodic function has an infinite number of frequencies, therefore the dimension of the attractor of the almost periodic dynamic system can be infinite. In this section we consider a new approach to the investigation of the attractor based on a study of sections. We define sections of the attractor as follows. Let $\mathcal{A}$ be the attractor of the semigroup $\{S(t)\}$ acting on $E \times H(\phi)$ generated by the family of processes $\{U_g(t, \tau)\}$, $g \in H(\phi)$, and let $\pi_1(\mathcal{A}) = \mathcal{A}_1$, $\pi_2(\mathcal{A}) = H(\phi)$. For each $g \in H(\phi)$ we define a set $\mathcal{A}(g) = \pi_2^{-1}(g)$. The set $\mathcal{A}(g)$ is called the section corresponding to the symbol g. It is clear that

$$\mathcal{A} = \bigcup_{g \in H(\phi)} \mathcal{A}(g).$$

We define also

$$\mathcal{A}_1(g) = \pi_1(\mathcal{A}(g)), \quad \mathcal{A}_1 = \bigcup_{g \in H(\phi)} \mathcal{A}_1(g).$$

The problem is to estimate the Hausdorff dimension $\dim_H \mathcal{A}(g)$ uniformly with respect to $g \in H(\phi)$ (or to estimate $\dim_H \mathcal{A}_1(g)$).

Let H be a Hilbert space, $\{U(t, \tau)\}$ a process acting on H generated by the non-autonomous equation

$$\partial_t u = -\mathcal{A}(u, t), \quad t \geq \tau, \quad u|_{t=\tau} = u_\tau \in H, \quad \tau \in \mathbf{R}. \tag{A6.1}$$

We consider a set X:

$$X = \{u(t) \mid t \in \mathbf{R}, u(t) \text{ is an arbitrary complete trajectory} \atop \text{of the problem (A6.1) bounded in } H\}. \tag{A6.2}$$

Definition A6.1.
A section of X at each time $s \in \mathbf{R}$ is the set

$$X(s) = \{u(s) \mid u(t) \text{ is an arbitrary bounded complete} \atop \text{trajectory of the problem (A6.1)}\}.$$

It is obvious that the following equality holds:

$$U(t, \tau)X(\tau) = X(t) \quad \forall t \geq \tau, \quad \tau \in \mathbf{R}. \tag{A6.3}$$

We assume that the process $\{U(t,\tau)\}$ is uniformly quasi-differentiable on X, i.e. there exists a family of bounded linear operators $\{U'(t,\tau,u)\}$, $u \in X$, $U'(t,\tau,u) : H \to H$ (quasi-derivatives) such that for any $t \geq \tau$, $\tau \in \mathbf{R}$,

$$\begin{aligned}
\|U(t,\tau)u_1 - U(t,\tau)u - U'(t,\tau,u)(u_1 - u)\| \\
\leq \gamma(t - \tau, \|u_1 - u\|)\|u_1 - u\|,
\end{aligned} \tag{A6.4}$$

where $u_1, u \in X(\tau)$, $\gamma(s,\xi) \to 0$ $(\xi \to 0)$ for each $s \geq 0$. The function $\gamma(s,\xi)$ doesn't depend on $u, u_1 \in X(\tau)$.

We assume that for $u_\tau \in X(\tau)$ the quasi-derivative $U'(t,\tau,u_\tau)$ is generated by the variational equation of the problem (A6.1):

$$\partial_t v = -A_u'(u,t)v, \quad v\,|_{t=\tau} = v_\tau, \quad t \geq \tau, \tag{A6.4'}$$

i.e. $U'(t,\tau,u_\tau)v_\tau = v(t)$, where $v(t)$ is the solution of (A6.4').

Let $Tr_d(L)$ be a d-trace of L. (See (A4.3), (A4.4)). Let us formulate the main result of this section, the analogue of Theorem A4.1.

Theorem A6.1.
Let $\{U(t,\tau)\}$ be a process uniformly quasi-differentiable on $[X]$ ($[\cdot]$ is closure in H) and $[X]$ be compact in H; X is defined in (A6.2). Let $X = \bigcup_{t \in \mathbf{R}} X(t)$ and

$$U(t,\tau)X(\tau) = X(t) \quad \forall t \geq \tau, \tau \in \mathbf{R}.$$

Let also

$$\|U'(t,\tau,u_\tau)\|_{\mathcal{L}(H,H)} \leq C(t - \tau) \quad \forall u_\tau \in X(\tau), t > \tau.$$

We define the value q_d:

$$q_d = \liminf_{T \to +\infty} \sup_{\tau \in \mathbf{R}} \sup_{u_\tau \in X_\tau} \left\{ -\frac{1}{T} \int_\tau^{\tau+T} Tr_d(\mathcal{A}_u'(u(s),s))\,ds \right\}, \tag{A6.5}$$

where $u(s) = U(s,\tau)u_\tau$. Then $q_d < 0$ implies $\dim_H X(t) \leq d$ for all $t \in \mathbf{R}$.

Here we omit the proof of Theorem A6.1. It is analogous to the proof of Theorem A4.1 but differs from it in some essential points.

Example A6.1.
We consider the Navier-Stokes system (A3.1):

$$\left.\begin{aligned}
\partial_t u = -Lu - B(u,u) + g(x,t), \\
u|_{t=\tau} = u_\tau, \quad t \geq \tau, \quad g \in H(\phi).
\end{aligned}\right\} \tag{A6.6}$$

The function $\phi(x,t)$ is almost periodic with values in H.

Under the assumption of Theorem A3.1 the problem (A6.6) generates the family of processes $\{U_g(t,\tau)\}$, $g \in H(\phi)$, acting on H. Theorem A3.1 implies that the family $\{U_g(t,\tau)\}$, $g \in H(\phi)$ possesses the compact uniform attractor $\mathcal{A}_{H(\phi)} = \mathcal{A}_1 \Subset H$. Let $\mathcal{A}_1(g)$ be a section of the attractor corresponding to a symbol $g \in H(\phi)$. One can use the notation (A6.3); then $X(0) = \mathcal{A}_1(g)$, $X(\tau) = \mathcal{A}_1(g^\tau)$, $g^\tau(t) = g(t+\tau)$, $X = \bigcup_{\tau \in \mathbf{R}} \mathcal{A}_1(g^\tau)$. The set X depends on $g \in H(\phi)$. Due to Theorem A3.1 the set X is precompact in H.

The following theorem holds.

Theorem A6.2.
For arbitrary $g \in H(\phi)$

$$\dim_H(\mathcal{A}_1(g)) \le \frac{C}{\nu^2}(M(\|\phi\|_{-1}^2))^{1/2} + 1, \qquad (A6.7)$$

where

$$M(\|\phi\|_{-1}^2) = \lim_{T \to +\infty} \frac{1}{T} \int_0^T \|\phi(s)\|_{-1}^2 \, ds.$$

Using standard methods one can show that the process $\{U_g(t,\tau)\}$ is uniformly quasi-differentiable on X. The variational equation is

$$\partial_t v = -Lv - B(u(t),v) - B(v,u(t)) \equiv -\mathcal{A}_u'(u)v. \qquad (A6.8)$$

The d-trace $Tr_d \mathcal{A}_u'(u)$ satisfies the following estimate (see Temam [2]):

$$-\frac{1}{t-\tau} \int_\tau^t Tr_d(\mathcal{A}_u'(U_g(s,\tau)u_\tau)) \, ds$$

$$\le -C_1 \nu d^2 + \frac{C_2}{\nu} \cdot \frac{\|u(\tau)\|^2}{t-\tau} + C_3 \cdot \frac{1}{\nu^3} \cdot \frac{1}{t-\tau} \left\{ \int_\tau^t \|g(s)\|_{-1}^2 \, ds \right\}.$$
$$(A6.9)$$

Finally note that $\|u(\tau)\| \le C_4$ for any $u(\tau) \in X(\tau)$ and

$$\frac{1}{T} \int_\tau^{\tau+T} \|g(s)\|_{-1}^2 \, ds \to M\|\phi\|_{-1}^2 \quad (T \to +\infty)$$

uniformly with respect to $\tau \in \mathbf{R}$ for any $g \in H(\phi)$ (see Bohr [1], Levitan & Zhikov [1]). The inequality (A6.9) implies (A6.7).

Example A6.2.   Time-dependent reaction-diffusion system with almost periodic terms

We consider the system (A1.9), where $\phi(x,t) \in C(\mathbf{R}, H)$ is an almost periodic function with values in $(L_2(\Omega))^N$. The function $f(u,x,t)$ satisfies conditions (A1.10) - (A1.13). We assume that $f(u,x,t)$ is an almost periodic function with values in the corresponding space L_ρ. The space L_ρ consists of all functions $\psi(u,x)$ having the finite norm

$$\|\psi(u,x)\|_\rho = \sup_{x \in \overline{\Omega}} \sup_{u \in \mathbf{R}} \left\{ \frac{|\psi(u,x)|}{|u|^{p-1}+1} + \frac{|\nabla_x \psi(u,x)|}{|u|+1} + \frac{|\psi'_u(u,x)|}{|u|^{p-2}+1} \right\}, n \geq 3.$$

We assume that all translations $\{f^h\}_{h \in \mathbf{R}}$, $f^h(u,x,t) = f(u,x,t+h)$, form a precompact set in $C(\mathbf{R}, L_\rho)$. Denote the corresponding hulls of functions ϕ and f by $H(\phi)$ and $H(f)$ respectively. The pair $(f(u,x,t), \phi(x,t)) \equiv P(t)$ is the time symbol of the problem (A1.9). Let $H(P)$ be a hull of the function P. Obviously $H(P) \subseteq H(f) \times H(\phi)$. It is easy to show that each $G(t) = (f_1(u,x,t), g(x,t)) \in H(P)$ satisfies the conditions (A1.10) - (A1.13). Therefore the problem (A1.9) generates the family of processes $\{U_G(t,\tau)\}$, $G \in H(P)$ acting on $(L_2(\Omega))^N$.

Theorem A6.3.

Let $f(u,x,t)$ and $g(x,t)$ be almost periodic functions (in the spaces defined above); then

(1) the family of processes $\{U_G(t,\tau)\}$, $G \in H(P)$, possesses the compact uniform attractor $\mathcal{A}_1 = \{u(0) \mid u(t)$ is an arbitrary bounded complete trajectory of the problem (A1.9) with any time symbol $G \in H(P)\}$,

(2) any section $\mathcal{A}_1(G)$ of the attractor has finite Hausdorff dimension:

$$\dim_H(\mathcal{A}_1(G)) \leq C,$$

where C doesn't depend on G and the expression for the constant C is similar to that in the autonomous case, i.e. when $f = f(u,x)$, $g = g(x)$. (See Babin & Vishik [1], Temam [2].)

The proof of Theorem A6.3 is analogous to the corresponding proof of Theorems A3.2 and A6.2.

Example A6.3.

The analogous theorem holds for the time-dependent hyperbolic equation (A1.20) where $f(u,t)$ is an almost-periodic function with values in the corresponding weighting space L_ρ and $g(x,t)$ is an almost periodic function with values in $L_2(\Omega)$. The corresponding family of processes have a uniform attractor $\mathcal{A}_1$ and all sections of this attractor have a finite dimension with uniform upper bound.

References

M. S. Agranovich & M. I. Vishik
[1] Elliptic problems containing a parameter and parabolic problems of general type, *Uspekhi Mat. Nauk*, 19, no. 3, 1964, pp. 53-160.

A. V. Babin & M. I. Vishik
[1] Attractors of evolution equations, *Moscow, Nauka*, 1989.
[2] On the unstable sets of evolution equations in a neighbourhood of critical points of a stationary curve, *Izvestiya Akad. Nauk, ser. mat.*, 51, no. 1, 1987, pp. 44-78.
[3] Regular attractors of semigroups and evolution equations, *J. Math. pures et appl.*, 62, 1983, pp. 441-491.
[4] Uniform finite-parameter asymptotics of solutions of nonlinear evolutionary equations, *J. Math. pures et appl.*, 68, 1989, 399-455.
[5] Attractors of evolutionary equations in partial derivatives and estimates of their dimension, *Uspekhi Mat. Nauk*, 38, no. 4, 1983, pp. 133-187.
[6] Unstable invariant sets of semigroups of nonlinear operators and their perturbations, *Uspekhi Mat. Nauk*, 41, no. 4, 1986, pp. 3-34.
[7] Spectral and stabilized asymptotic behaviour of solutions of nonlinear evolutionary equations, *Uspekhi Mat. Nauk*, 43, no. 5, 1988, pp. 99-132.
[8] Maximal attractors of a semigroup, corresponding to evolutionary differential equations, *Mat. Sbornik*, 126, no. 3, 1985, pp. 397-419.
[9] Attractors of parabolic and hyperbolic equations. Properties of their compactness and attraction, *Vestnik of Moscow University, mat. and mechanics ser.*, no. 3, 1988, pp. 70-72.

O. V. Besov, V. P. Ilyin & S. M. Nikolsky
[1] Integral representations of functions and embedding theorems, *Moscow, Nauka*, 1975.

H. Bohr
[1] Fastperiodische funktionen, *Berlin, Verlag von Julius Springer*, 1932.

P. Constantin, C. Foias & R. Temam
[1] Attractors representing turbulent flows, *Mem. Amer. Math. Soc.*, 53, no. 314, 1985.

C. M. Dafermos
[1] Asymptotic behaviour of solutions of evolution equations, *Nonlinear evolution equations, New York Academic Press*, 1978, pp.103-123.
[2] Semi-flows associated with compact and almost uniform processes, *Math. Systems Theory*, 8, 1974, pp. 142-9.
[3] Almost periodic processes and almost periodic solutions of evolution equations, *Proceedings of a University of Florida International Symposium, New York Academic Press*, 1977, pp. 43-57.

C. Foias & C. Guillopé
[1] On the behaviour of the solutions of the Navier-Stokes equations lying on invariant manifolds, *J. Diff. Eq.*, 61, 1986, pp. 128-48.

C. Foias & J. Saut
[1] Asymptotic behaviour at $t \to +\infty$ of solutions of Navier-Stokes equations and nonlinear spectral manifolds, *Indiana Univ. Math. J.*, 3, 1984, pp. 459-77.

J. K. Hale
[1] Asymptotic behaviour of dissipative systems, *Mathematical Surveys and Monographs 25, Amer. Math. Soc., Providence, RI*, 1988.
[2] Asymptotic behaviour and dynamics in infinite dimensions, *Res. Notes in Math.*, 132, 1985, pp. 1-42.

J. K. Hale & G. Raugel
[1] A damped hyperbolic equation on thin domain, *Centre for Dynamical Systems and Nonlinear Studies*, 1989.

A. Haraux
[1] Asymptotics for some nonlinear O.D.E. of the second order, *Université Pierre et Marie Curie, Centre National de la Recherche Scientifique, Publication de Laboratoire d'Analyse Numérique*, N 85015, 1985.
[2] Asymptotics for some nonlinear hyperbolic equations with a one-dimensional set of rest points, *Bol. Soc. Bras. Mat.*, 17, no. 2, 1986, pp. 51-65.
[3] Systèmes dynamiques dissipatifs et applications, *Paris, Milan, Barcelona, Rome, Masson*, 1991.
[4] Two remarks on dissipative hyperbolic problems, *Coll. France Seminar, Pitman Res. Notes Math.*, 6, 1985.

M. Hirsh, C. Pugh & M. Shub

[1] Invariant manifolds, *Lect. Notes in Math. vol. 583, Springer-Verlag,* 1977.

Y. Ilyashenko

[1] Weak squeezing systems and attractors of Galerkin approximations of Navier-Stokes systems, *Uspekhi Mat. Nauk,* 36, no. 3, 1981, pp. 243-4.

[2] Weak squeezing systems and attractors of Galerkin approximations of Navier-Stokes system on two-dimensional torus, *Uspekhi of Mechanics,* no. 1, 1982, pp. 31-63.

O. A. Ladyzhenskaya

[1] Mathematical problems in the dynamics of a viscous incompressible liquid, *Moscow, Nauka,* 1970.

[2] The minimal global attractors for Navier-Stokes equations and other partial differential equations, *Uspekhi Mat. Nauk,* 42, no. 6, 1987, pp. 25-60.

B. Levitan & V. Zhykov

[1] Almost-periodic functions and differential equations, *Moscow University Press,* 1978.

J.-L. Lions

[1] Quelques méthodes de résolution des problèmes aux limites non linéaires, *Dunod, Gauthier-Villars, Paris,* 1969.

J. E. Marsden, M. McCracken

[1] The Hopf bifurcation and its applications, *New York, Springer-Verlag,* 1976.

X. Mora, J. Sola Morales

[1] The singular limit dynamics of semilinear damped wave equation, *Département de Mathématiques Universitat Autonoma de Barcelona,* 1987.

S. M. Nikolsky

[1] Approximation of functions of several variables and embedding theorems, *Moscow, Nauka,* 1969.

V. A. Pliss

[1] Principle of reduction in the theory of stability, *Izvestiya Akad. Nauk, ser. mat.,* 28, 1964, pp. 1297-1324.

G. R. Sell

[1] Nonautonomous differential equations and topological dynamics I, II, *Amer. Math. Soc.,* 127, 1967, pp. 241-62, 263-83.

[2] Topological dynamics and ordinary differential equations, *London, Van Nostrand Reinhold,* 1971.

154   References

M. Y. Skvortsov

[1] Maximal attractor of semigroup, corresponding to the first boundary value problem for parabolic equation with singular perturbation, *Uspekhi Mat. Nauk*, 42, no. 4, 1987, pp. 243-4.

[2] Attractors of singular perturbed dynamic systems, *Uspekhi Mat. Nauk*, 42, no. 4, 1987, p. 154.

[3] Attractors of singular perturbed dynamic systems, *Depon. VINITI - B86*, pp. 1-36.

M. Y. Skvortsov & M. I. Vishik

[1] Attractor of singular perturbed parabolic equation and asymptotic behaviour of its elements, In: *Advances in Soviet Mathematics*, vol. 10, *Properties of Global Attractors in Partial Differential Equations*, eds. A. V. Babin and M. I. Vishik, Amer. Math. Soc., Providence, R.I., 1992.

R. Temam

[1] On the theory and numerical analysis of the Navier-Stokes equations, *Springer-Verlag*, 1974.

[2] Infinite-dimensional dynamical systems in mechanics and physics, *Springer-Verlag*, 1988.

A. N. Tichonov

[1] Systems of differential equations with small parameters, *Mat. Sbornik*, 31, no. 3, 1952, pp. 375-86.

[2] Systems of differential equations with small parameters, *Mat. Sbornik*, 27, no. 1, 1950, pp.147-56.

J. Wells

[1] Invariant manifolds of nonlinear operators, *Pacific J. of Math.*, 62, no. 1, 1976, pp. 285-93.

M. I. Vishik & A. V. Fursikov

[1] Mathematical problems of statistical hydromechanics, *Domtrecht, Boston, London, Kluwer Academic Publishers*, 1987.

M. I. Vishik & L. A. Lusternik

[1] The regular degeneration and boundary layer for linear differential equations with small parameter, *Uspekhi Mat. Nauk*, 12, no. 5, 1957, pp. 3-122.

[2] Solution of some problems concerning perturbations in the case of matrices and self-adjoint and nonself-adjoint differential equations, *Uspekhi Mat. Nauk*, 15, no. 3, 1960, pp. 3-80.

M. I. Vishik & M. Y. Skvortsov

[1] Attractor of singular perturbed parabolic equations and asymptotic behaviour of its elements, *Mat. Sbornik*, 182, no. 12, 1991, pp.

1769-85.

[2] The asymptotics of trajectories lying on the attractor of singularly perturbed parabolic equation, *Vestnik Moscow University*, ser. I, *Mat. Mechanics*, no. 6, 1991, pp. 11-16.

M. I. Vishik & V. Y. Skvortsov

[1] The asymptotics of solutions of reaction-diffusion equations with a small parameter, *to appear*.

[2] Stabilized asymptotics of solutions of parabolic systems depending on a parameter, *Uspekhi Mat. Nauk*, 45, no. 4 (274), 1990, pp. 134-5.

Printed in the United States
By Bookmasters